In-Process Quality Control for Manufacturing

INDUSTRIAL ENGINEERING

A Series of Reference Books and Textbooks

Editor

WILBUR MEIER, JR.

Dean, College of Engineering
The Pennsylvania State University
University Park, Pennsylvania

Additional Volumes in Preparation

In-Process
Quality Control
for Manufacturing

WILLIAM E. BARKMAN

Martin Marietta Systems, Inc.
Oak Ridge, Tennessee

MARCEL DEKKER, INC.   New York and Basel

Library of Congress Cataloging-in-Publication Data

Barkman, W. E.
  In-process quality control for manufacturing.

  (Industrial engineering ; v. 16)
  Bibliography: p.
  Includes index.
  1. Process control — Statistical methods. 2. Flexible
manufacturing systems.   I. Title.   II. Series.
TS156.8.B37   1989   670.42   88-36328
ISBN 0-8247-8054-X (alk. paper)

This book is printed on acid-free paper.

MARCEL DEKKER, INC.
270 Madison Avenue, New York, New York 10016

Current printing (last digit):
10 9 8 7 6 5 4 3 2 1

PRINTED IN THE UNITED STATES OF AMERICA

Preface

The idea that concerns about quality control are or should be an inherent part of all manufacturing operations is a readily understood concept. In addition, this philosophy is currently receiving a large amount of attention in the manufacturing community due to the increasing challenge to produce products that are competitive in price and utility in the world marketplace. In almost all instances, this means that today's manufacturing operations can be characterized by the fact that it is undesirable to make products that are not fit for their intended use (scrap). In order to accommodate this situation, a variety of solutions may be proposed to control the quality of the finished workpieces. Rigorous testing of each individual part can be employed at the end of the manufacturing cycle to establish the quality level of the process output (within the error limits of the inspection equipment). Then, through a selection process, it is possible to assure that the quality level of the products that the customer receives will meet the agreed upon requirements within a specific level of confidence. (For instance, it could be estimated ahead of time that 99.99% of all of the products that are shipped to a customer will meet specific performance requirements.) This may be an acceptable mode of operation as long as there is a sufficient market for the lower quality products and the customer with the higher quality requirements is willing to accept the extra product certification expenses that are encountered. However, this sorting procedure does nothing to control the actual quality of the manufacturing operation—it only establishes a historical data base that describes the results of the manufacturing process. An alternative approach for improving the quality of a process's output is to employ manufacturing techniques that depend on statistical quality control, as demonstrated by Demming and others.

I obviously support the concept of applying statistical quality control to manufacturing applications. However, it is recognized that the term *statistics* is often intimidating to people, so that rather than leaning heavily on theoretical analysis, this text deals more with application examples and techniques for intelligently applying statistical methods to the manufacturing world. Common sense blended with the appropriate advanced technology is the goal that is pursued. In addition, the notion of determinism in the manufacturing world is discussed. While this term may sound exotic, it is really only an extension of common sense, which serves to provide a label for a method of operation that is frequently applied piecemeal without being integrated throughout an entire manufacturing facility.

Webster's Collegiate Dictionary defines determinism as "a doctrine that acts of will, occurrences in nature, or social or psychological phenomena are determined by antecedent causes." For the purposes of this text, it will be assumed that the social implications of determinism are ignored and that only the "action-reaction" relationship in nature is being considered. Belief in a "cause-and-effect" environment leads to the idea that one means of attaining control of a manufacturing process is to monitor and/or control the key parameters that are most significant in establishing the quality of the finished product. By definition, this text assumes that all manufacturing processes are deterministic. However, it is also recognized that a major difficulty that can be encountered is the identification and isolation of the few critical parameters that have a major effect upon the process quality (as opposed to a myriad of relatively unimportant factors). An application of deterministic manufacturing is one in which these critical parameters are identified and the appropriate sensors are available for parameter monitoring.

This text was written to meet the rapidly changing needs of the practicing engineer or engineering manager who is involved in the manufacturing environment. In addition, it should be very useful to the academician who is concerned with addressing the real-world difficulties of the manufacturing community. The information that is included in the following pages will be valuable to those who have an intimate understanding of their own manufacturing problems while perhaps forgetting much of what was covered in their college courses. A background in statistics, theoretical mathematics, or computer science is not required to obtain a complete understanding of the manufacturing approach that is presented. However, as is true in most situations, a high degree of common sense is quite helpful. With the exception of the two chapters that concentrate on statistical methodology, there is a lack of equations and formulas within the text. Instead of presenting theoretical information to be memorized, the objective is to allow the reader to gain an intuitive understanding of the precepts of deterministic manufacturing which is based on a sound statistical footing. The intention is to enable the individual working in the real world of production schedules and increasingly complex and demanding tolerance requirements to develop a manufacturing philosophy which builds quality into the workpiece rather than trying to inspect quality in after the fact.

The information presented here ranges from a brief definition of what constitutes quality and a manufacturing environment to a detailed description of the entities that make up a computer-controlled deterministic manufacturing operation. The concepts of process parameters and computer-controlled manufacturing are explained and specific information is provided concerning types of sensors that can be used to provide the necessary process description data to the computer control system. The chapters on computer networks and system software describe how the process control data is sent from one computer to another and how the data is manipulated to achieve the desired control functions. Additional information is presented to show how the deterministic manufacturing concept blends into flexible manufacturing systems and the factory of the future. It should also be recognized that while this material is concerned primarily with piece-part manufacturing, the ideas presented are equally applicable to the process control industries.

The detailed information necessary to completely define and implement a computer-controlled deterministic manufacturing system is quite extensive, varies considerably for different tasks, and often requires the assistance of experts from multiple technical disciplines. In addition, it is not always necessary to employ a completely automated system to achieve the benefits of deterministic manufacturing methods. For some applications, a manual or semiautomatic system may be a more appropriate means to implement these techniques. Because of this wide variety of circumstances as well as the author's specific manufacturing background, the application examples given are concerned primarily with the metal working industries. Providing detailed information for all manufacturing operations is beyond the scope of this text. However, the material in these pages will provide a firm foundation to those persons wishing to upgrade their manufacturing operations through the application of the deterministic manufacturing approach.

William E. Barkman

Contents

CHAPTER 1

Introduction

Building quality control into manufacturing operations is almost a uniform goal throughout industry and it is judged to be so important that it is essentially a sacred tenet. Certainly the pursuit of improved products through the use of computer control or other appropriate means is an admirable goal. In addition, the search for this goodness is in tune with our apparent need to ever advance on the frontiers of technology, as well as being relatively attainable, unlike other holy grails. But what does quality really mean in the manufacturing environment aside from the terms high, low, medium, etc., that are often discussed? From a technical standpoint, quality implies a degree of excellence which causes a product to be superior to others in performing its intended function. In addition, less tangible factors such as appearance and feel or even market image may be used in the evaluation of an item's quality. Product specifications are one means of comparing the quality of different entities, although this can be deceptive in some instances.

While cost effective quality control is required to maintain a competitive position in the marketplace, in general terms this objective can be achieved through one of three basic approaches. One option that can be pursued is to strive to never make a defective product, another avenue is to attempt to ensure that no faulty products are shipped to the customer, and the third method is a combination of the first two. Each of these approaches depends to some degree on a combination

1

of product inspection and statistical evaluation of the inspection results. If every thing was perfect, there would no need for statistics or statisticians since they thrive on errors. However, since we live in an imperfect world it is necessary to employ this tool to our advantage. (Some people might term statistics a "necessary evil.") This introduction of statistics into the product quality picture leads to the use of technologies such as statistical process control (SPC) and statistical quality control (SQC) which appear frequently in the current trade journal literature.

It has been said that many people use statistics like a drunk uses a lamp post—for support rather than illumination. While this may be true, it does not reflect on the validity of statistics, it merely recognizes the misuse of a very powerful tool. Statistics can be legitimately used to support a particular theory, and the option for gaining understanding about a situation through the use of statistical techniques should not be ignored.

In a similar fashion, the term determinism is often misunderstood. Manufacturing processes are deterministic, which means that there is a specific reason or set of reasons for the condition of the quality of the output of a process. (People are not deterministic in the mechanical sense associated with manufacturing operations.) However, understanding the statistical characteristics of a group of parts does not determine the condition of an individual part, it only defines the probability that an individual workpiece will exhibit certain features or parametric values.

Obviously, different manufacturing processes are influenced in unique ways by their various process parameters. However, if the key process variables are maintained in a constant state, then the product quality will be unchanged over time. In actual practice, it is frequently difficult or impractical to attempt to maintain a process parameter at an exact value. Instead, some normal range of values exists for the parameter when everything is working properly. In this case, there will be no more variation in the product quality than that caused by the normal variation in the key process parameters. If there is an unusual shift in the quality of the product, then there must be an antecedent cause directly related to the condition of one or more of the process parameters. "Process shifts" don't just happen at random or without cause. When a shift does occur, there is a specific reason for this occurrence which is directly related to the condition of the process parameters.

Often, three major factors are involved in a perturbation in a given process. These factors are the quality of the input material, the quality of the fabrication equipment, and the presence of manual operations. For example, assume that company BB manufactures ball bearings which are manually inspected for size using hand-held micrometers. These ball bearings are sold to company LS where the bearings are incorporated into leadscrews for moving machine slides. In addition, suppose that company RB uses leadscrews from company LS to manufacture a robot positioning system for loading electrical components onto printed circuit boards. Assume further that, unknown to anyone outside of company BB, a new employee is assigned to manufacture the ball bearings. One part of the ball bearing manufacturing process is the inspection of the quality of the ball bearings and the manual adjustment of the fabrication process to obtain the desired process parameters. This new employee performs the manual measurements without a clear understanding of the way in which the manual gages are to be used or how to compensate for variations among different gages. As a result defective ball bearings are shipped to company LS for use in their leadscrews. Because of a lack of adequate testing of incoming material, company LS uses the faulty ball bearings in leadscrew assemblies which are then sent to company RB for use in their robot positioning systems. After a period of time the faulty ball bearings cause early failure in the leadscrews and the robots begin to produce poor quality printed circuit boards which are not discovered until a number of the units have already been shipped to customers. The factor involved in the process shift for company RB is the quality of their own fabrication equipment, which is influenced by the incorrect use of manual gages at company BB. Testing of incoming product qualty by company LS could also have helped. Automation of the ball bearing testing and process adjustment operations by company BB would be one way to avoid problems in product quality, while company LS could also provide an automatic system to test the quality of their incoming products. All three companies would benefit from the incorporation of statistical quality control techniques in a real-time system to monitor the status of their manufacturing operations.

Automatic test equipment for the evaluation of incoming and outgoing product quality is prevalent in much of industry today. One example of a device that is used to provide dimensional information about a product is the Cartesian coordinate measuring machine (CMM) shown in Figure 1.1. This machine is controlled by a computer numerical control (CNC) system which can be programmed to perform the complete dimensional inspection task without operator intervention. After a part is loaded in a general location on the worktable of this machine, the control system directs the gage to establish the location of the part in the machine measurement space through the use of the inspection probe. Then, the system determines the tilt of the part datum with respect to the machine axes and adjusts the measurement data accordingly. This type of automatic datum correction operation is particularly important for dimensional inspection systems that employ robot handling. Since the precision of most robots is not equal to the dimensional tolerances required on the workpieces, the part alignment errors will be interpreted as product dimensional errors unless the data correction sequence is utilized.

In contrast, the use of in-process measurements of product characteristics for the control of process quality is rather limited. In some instances, this is because of the slowdown that this type of activity can introduce into the manufacturing cycle. In other cases, this technique is not utilized either because it is not

Figure 1.1 Computer numerical controlled coordinate measuring machine (courtesy of Martin Marietta Energy Systems, Inc.).

understood or accepted or because the knowledge required for its application is unavailable. Parity checking in a computer system is an example of the use of a limited degree of real-time monitoring, while the redundant computer systems on aircraft demonstrate the more extensive use of process monitoring.

An alternative approach to the utilizaticn of direct, in-process product measurements for estimating product quality is the concept of "process metrology" [1]. This method of operation is based on the theory that the determinism doctrine is valid in the manufacturing world. The deterministic manufacturing approach is founded on the precept that there is a cause and effect relationship in existence in manufacturing operations (it is recognized that this is not true on an atomic level, as shown by quantum theory). This means that process variables can be utilized in a statistical fashion to estimate the quality of a process. In addition, this prediction of the quality level of the end product can be accomplished without having to resort to in-process inspection of actual product attributes nor is it necessary to wait until the manufacturing operations are completely finished. An obvious advantage of this mode of operation is that shifts in the manufacturing process characteristics can be detected immediately. This timely process quality information allows corrective action to be taken prior to making faulty products rather than after it is too late to avoid producing some amount of scrap. In addition, the expense of postprocess inspection is reduced significantly. This occurs because once the critical process attributes are employed to describe the manufacturing operation, the postprocess inspection steps are required only to provide a statistical verification of the continued validity of the system model or to resolve a question concerning a product with "marginal" characteristics.

The continuous process industries, which deal with products such as petroleum or chemicals, are in direct contrast to piece-part manufacturing activities in that the measurement of the process variables is common place. In addition, it is relatively difficult to make a direct in-process measurement of the product attributes for a continuous process except through the use of spot sampling of the product stream. One simplistic example of the use of the key process parameters for monitoring and/or controlling an operation is the process of baking a cake. In this situation, a number of ingredients are mixed together and placed into an oven. To achieve the desired result, it is necessary to leave the cake mix in the oven for a specific time while keeping the temperature of the interior of the oven within a certain temperature range. If the quality of the input ingredients is controlled, a time–temperature relationship must be maintained to control the quality of the finished product. A lower than normal oven temperature can be offset, to a certain extent, by increasing the baking time. However, an unplanned significant increase in time or temperature will result in the cake being burned, and it is not necessary to wait until the end of the normal baking cycle to taste the cake and evaluate the results of this shift in process parameters. Unfortunately, most manufacturing processes are not as simple as baking a cake and the execution of a computer model of the system operation may be required to make an accurate prediction of the final quality level of the process output. In addition, the key process characteristics may not be as easy to measure as baking time and oven temperature. Examples of the types of parameters that might be required to employ the process metrology approach in manufacturing operations include machine and component temperature profiles, servo system signals, vibration frequencies and amplitudes, cycle times, tool wear, component deflection or run out, and so on.

It is probably apparent that highly predictable or automated systems are required to achieve a viable deterministic manufacturing operation. This is because these systems provide the consistency of operation required to obtain a high level of quality as well as permit a useful statistical estimation of the process quality. In addition, automated systems are often able to perform tasks that are beyond the abilities of human operators. The generation of a contoured part on a lathe is one example of this increase in performance capability. A CNC lathe can easily generate a wide variety of workpiece shapes, but a human operator cannot even begin to match this capability using a manual machine. Another predictability factor that must be considered is that the input material to a process must be uniform. Otherwise, provisions must exist to detect and accommodate the expected variations in this process parameter. Finally, the process must be sufficiently understood and characterized by the measured variables so that a sound statistical analysis of the operation can be achieved in real time. Otherwise, any process adjustments are likely to increase the magnitude of an existing problem.

Chapter 1 begins the process monitoring section of the text with a user-friendly discussion of the statistical approach to quality control. This is accomplished through a low-key review of the basic elements of probability and the integration of this material into the practical world of the engineer who is associated with maufacturing operations. Easily understood examples are given to show how the various statistical tools can be applied in manufacturing processes. Chapter 2 continues with a discussion concerning the application of statistical techniques to manufacturing operations and covers deterministic manufacturing methods in depth. Error sources, process parameters, and operational strategies are discussed and actual industrial applications are described. Chapter 3 extends the statistical material presented in Chapter 1 and covers random sampling, confidence limits, control charts, and error budgets. Chapter 4 expands the process monitoring discussion and presents the variety of options available. The need to restrict the monitored parameters to a minimum number while maintaining a sufficiently accurate characterization of the process is discussed. Different approaches that may be taken to develop the appropriate process model for use with the monitoring system are presented. The material contained in Chapter 5 completes the first phase of the text by discussing the multiple types of sensors that are available to monitor the critical process parameters for various manufacturing operations.

The key to the utilization of manufacturing process measurements for obtaining a system that is capable of controlling the process, without depending exclusively on postprocess measurements, is the use of computer-control technology. Of course, one of the first laws of computer systems is "garbage in, garbage out." This means that the mere existence of computer hardware and software in a given operation is not a panacea since any computer system is able to perform only those tasks defined by its operating software. Chapter 6 begins the computer-related portion of the text by discussing computer-controlled manufacturing, including the systems commonly known as NC, CNC, DNC, CAM, and AI. Chapter 7 deals with computer networks and computer communications and some of the complications that can arise when computer interconnections are encountered. Chapter 8 is devoted to a subject that is frequently the most underestimated area in computer controls—system software. Some of the types of the wide variety of software that are likely to be encountered in the application of computer systems to manufacturing operations are discussed. Chapter 9 covers the critical subject of the integration of manufacturing systems that employ computers, including operations requirements that occur during system start-up. Chapter 10 discusses some aspects of the much heralded factory of the future that is beginning to take shape but that will never totally arrive since there is always "one more" improvement that could be implemented.

Reference

1. J. S. Simpson, *Metrology of the Future and the NBS Manufacturing Research Facility*, U.S. National Bureau of Standards, Washington, D.C.

CHAPTER 2

Statistical Descriptions of Process Quality

Introduction

Quality was described in the introduction as a measure of goodness that relates to the intended use of a product and the expectations customers have concerning this product. However, it may not be obvious to many people what part statistics can play in the search for quality. Most manufacturing personnel recognize that statistical "mumbo jumbo" can be employed to describe some facets of a group of data such as averages, standard deviations or variances. However, the calculation of this "mystical number" is not the sole benefit of utilizing statistical methodology. The advantage of the careful application of statistical analysis techniques to a set of data is that the result is a descriptive parameter, or figure of merit, that is well defined mathematically and is relatively free of personal bias (although statistical procedures can be abused just like other data analysis approaches). This statistically determined performance parameter also provides a mechanism for obtaining an unbiased estimate of a process's ability to achieve a particular level of success in meeting the desired attributes. When working with a statistically stable process (a process which is predictable), this approach allows valid comparisons to be made of different operating conditions so that the system's performance can be optimized in terms of the specific desired product characteristics. The careful use of statistics permits opinions to be replaced with facts.

Manufacturing operations, for a particular batch or group of workpieces, are frequently characterized by a series of repetitive, sequential and simultaneous events that are performed in the same order from one part to the next. In addition, each of the workpieces generally experiences approximately the same environmental conditions during the manufacturing operations. Because of this similarity in fabrication cycles and the relative uniformity of the operating conditions, the end result is a group of finished workpieces which exhibit similar characteristics or attributes. Some examples of manufacturing activities typical of this type of situation include making a batch of 25,000 light bulbs, manufacturing automobiles, filling each of 10,000 bags with 15 marbles, or building VCR cameras.

One phenomena associated with this class of repetitive manufacturing operations is that even though an attempt is made to maintain constant manufacturing conditions at all times, the output products will exhibit attributes which display a certain variation from one part to the next. This is the normal variability in product condition or quality that is associated with a statistically stable process, and it is a natural result of the statistical probabilities that are inherent in manufacturing processes. Since none of the actions that make up the manufacturing operation occur in *exactly* the same fashion from one cycle to the next, it is not unexpected that the results of the process will also be somewhat unpredictable. While it is relatively easy to characterize the range of values between which a particular process characteristic will fall, it is not likely that the *exact* value can be successfully predicted. The study of these variations in the characteristics of a manufacturing process, which are traceable to specific events within the manufacturing process, begins with an area of statistics called probability theory.

If we consider a hypothetical problem such as the task of filling each of 10,000 bags with 15 marbles, it becomes apparent that it is possible to end up with a minimum of zero marbles in a bag, the maximum number of marbles that the bag can hold (assume 20), or any number between 0 and 20. In other words, there are 21 possible events that could occur: e(1), e(2),. . ., e (21). In a statistical sense, each of these events is a random event since the occurrence of one situation does not influence the occurrence of the others (assuming the nonexistence of a pathological condition such as the machinery becoming jammed so that no marbles are available for insertion into the containers). An example of a nonrandom event is the relationship between peoples' height and weight. That is, tall people tend to weigh more than short people. On the other hand, hair color is generally not a function of height.

Returning to the previous example, it can be assumed that f(1),f(2),. . .,(f)21 are the respective fractions of occurrences for the events e(i), or the frequency with which bags are filled with from 0 to 20 marbles. Then, the sum of the f(i) values is equal to 1 since the bags must end up with from 0 to 20 marbles in all cases (100% of the time). As the operation continues, assuming the process is in equilibrium, the f(i) values will stabilize and be equal to the probability of occurrence, p(i), for each of the events, e(i). In fact, the probabilities of each of the events, e(i), occurring will be distributed over the range from 0 to 100 percent in a fashion that is descriptive of the quality of the bag-filling process. The ideal situation is for each bag to have exactly 15 marbles or p(16) = 1. In actual cases, it is more likely that p(16) will be some value that is less than one and the sum of the other p(i) will be greater than zero. However, from a quality control standpoint it is desirable to have p(16) be as large as is practical.

Table 2.1 shows hypothetical data on the number of occurrences for each of the possible events for two-bag-filling systems in the imaginary marble-manufacturing operation. This information represents test results that describe

Table 2.1　Probability of Packaging Numbers of Marbles in Bags

Event (marbles/bag)	Frequency of Occurrence in 10,000 Bags	
	Process A	Process B
0	0	100
1	0	0
2	10	0
3	10	0
4	20	0
5	30	0
6	40	0
7	50	0
8	70	0
9	100	0
10	110	1
11	150	2
12	200	5
13	500	10
14	1,000	25
15	5,740	9,814
16	1,000	25
17	500	10
18	200	5
19	150	2
20	120	1

the number of marbles that were inserted into each bag when the operations were allowed to run undisturbed for an extended period of time. This data defines the frequency with which the two different processes filled the bags with each of the possible values over a period of 10,000 cycles. Dividing the number of times that the bags were filled with each of the possible numbers of marbles by the total number of trials gives the probabilities, p(i), for each of the events, e(i). This information may also be presented in graphical form through the use of a bar chart. This type of plot is called a frequency histogram. It consists of a plot of the total times that a particular event occurred within the testing period.

Alternatively, the relative frequency or probability of occurrence of each of the events also may be plotted. The only difference in these two types of graphical displays is the scale used on the ordinate axis. In addition, this type of plot can be used with ranges of data. For instances, each bar on the histogram could be scaled to depict groups of three values such as 0-2, 3-5, 6-8, etc. The key requirement is that the area contained within the histogram rectangles is proportional to the frequency of the observations within the particular interval. This does not require that the size of the intervals be the same, although they are usually presented in this manner for convenience. Also, as the number of process samples is increased the frequency histogram becomes smoother and more continuous in appearance. As might be expected, this same effect is observed as the size of the increments plotted along the abscissa is decreased. The cumulative result is that a continuous curve can be drawn between the centers of the tops of the histogram bars as a close approximation of the actual process. This curve is a useful means of representing the characteristics of a process but its origin must be remembered so that unrealistic inferences are not attempted. For example, although a continuous curve might imply that the average bag coming out of the process would contain 16.397 marbles, this is obviously an inappropriate interpretation of the statistical information (assuming the marbles remain intact during the loading cycle).

Figures 2.1a and 2.1b are graphical displays called histograms that show the relationship between the individual events, e(i), and the probability of these event's occurrence, p(i), for the two bag-filling operations. These plots are useful because they depict the probability of occurrence of a random variable distributed over its range of possible values; however, they do not indicate whether or not a process is in a state of statistical control. (A variable is some characteristic of an operation—in this case the number of marbles per bag.) One of the bag-filling systems, process A, has a nonsymmetrical distribution of probabilities. This type of distribution can be caused by a process with a lot of scatter plus the physical limitations of the bag—it can only hold 20 marbles. If larger bags were used, the distribution for process A would probably range from about 2 to 28 marbles per bag. However, with the smaller bags any marbles over 20 simply will not fit and the distribution curve is truncated. Process B exhibits much better control over the number of marbles per bag except for the occasional occurrence of empty bags. Because of the otherwise excellent characteristics of process B, it might be expected that the occasional process errors that result in empty marble bags are caused by some factor unassociated with the actual operation of filling the bags (prehaps a phenomena exists that allows some of the bags to bypass the bag-filling station).

While Figure 2.1 displays the probability of occurrence of each of the e(i) events, it also permits the calculation of the probability that an event will occur that is larger or smaller than a particular value. As an example, the probability of the

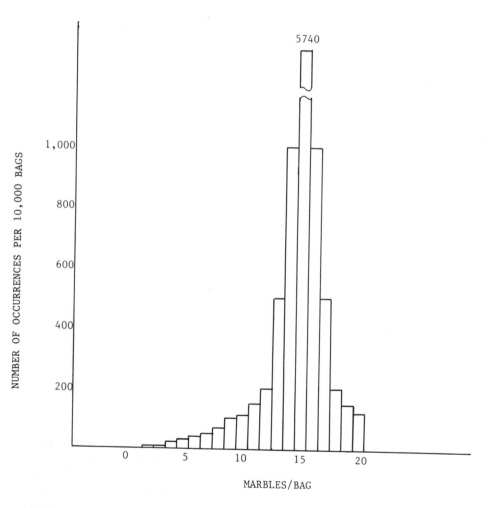

(A)

Figure 2.1 Bar charts of Marble packing probability for processes A (above) and B (next page).

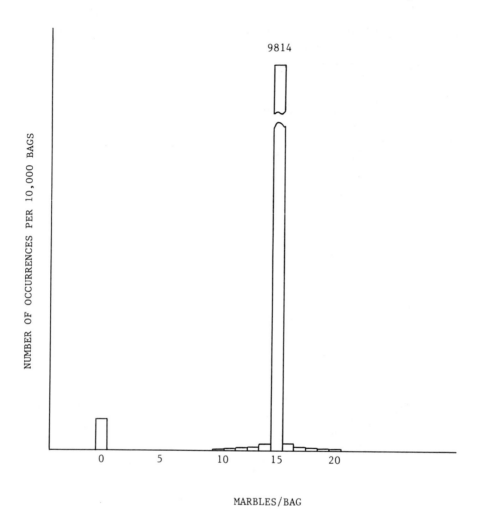

(B)

Figure 2.1   *(continued)*

system producing bags with more than 15 balls in them is represented by the area included in the histogram rectangles for events greater than 15 balls per bag. Figure 2.2 shows another way of graphically presenting this information. In this case, the data from Table 2.1 is plotted in a manner that is called a cumulative probability distribution. This is simply another way of presenting the previous data in a different form. This probability distribution is merely the summation of the individual event probabilities, p(i), when moving from event e(1) to e(21) and provides information on the probability of having at least a certain number of marbles in a particular bag. (The probability of having less than a certain number of balls in a bag can be calculated by taking the appropriate number from the cumulative probability distribution and subtracting it from one). As should be expected, the plots for the two processes are significantly different due to the large amount of scatter associated with the one process. Process A has an increasing possibility of event occurrence beginning with 2 marbles per bag. Process B has a probability of the occurrence of empty bags but there is no further variation from the desired value until the possibility of encountering 10 marbles per bag is considered.

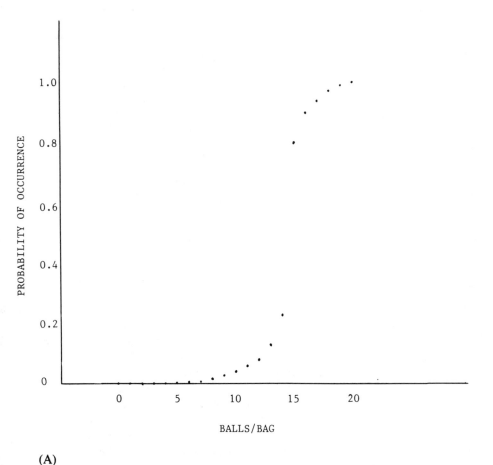

(A)

Figure 2.2 Cumulative probability distributions for processes A (above) and B (next page).

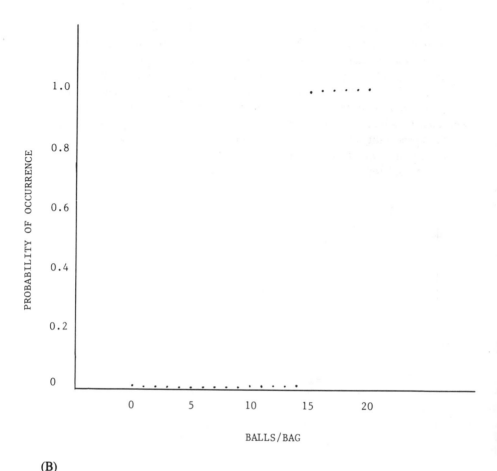

(B)

Figure 2.2 *(continued)*

An alternative approach for describing these and other processes is to use statistics such as the maximum, minimum, range, mean, median, variance, and standard deviation. The maximum and minimum are the largest and smallest quantities respectively, that occur within a group and the range is the difference between these two values. The mean or average value is that number which describes the middle of a data set in that the sum of the values occurring above the mean value is the same as the sum of the values occurring below it. The median value is the point at which the number of occurrences above and below a value are identical. Often the mean and median are similar numbers. While the mean and the median are useful statistics, they usually do not adequately describe a process. In the bag of marbles example, both processes A and B have similar means but the actual processes are quite different in terms of their overall capabilities. The standard deviation and variance are statistics that are used to provide additional information about the characteristics of a process. These statistics describe the variability among the population being studied and readily highlight the quality level differences in processes such as A and B.

Interest in statistics for describing the quality of a manufacturing operation is frequently centered around questions such as "What is the probability that this fabrication process will produce parts that will meet the design tolerence?" or "What is the probability of making an unacceptable workpiece?" Another topic of frequent interest involves conditional probability. This means that given the occurrence of one event, what is the probability that another event will also happen. In the above marble packaging example we might be interested in the probability that an inspection station would detect a bag with 10 marbles in it if the bag-filling machine is assumed to have made an error. This type of situation also leads to the topics of random sampling and control charts since it is desirable to provide the inspection operation or quality assurance system with effective tools for rapidly detecting the existence of problems within a particular manufacturing process.

The following sections of this chapter as well as a later chapter on the statistical analysis of manufacturing processes discuss these areas in more detail and give more specific instances of practical applications. It may be that the depth of the discussion of statistics or the use of equations (essentially limited to these two chapters) involves excessive detail for some readers. If this is the case, it is suggested that the beginning of each section of these chapters be read but that excessive time not be spent worrying about an in-depth understanding of the statistical details. However, regardless of the level of detail that is comfortable for each individual there is one point that should be remembered—the use of statistics is likely to be misleading (or worse) unless the data was collected from a statistically stable process. (This type of process is one which is predictably random as opposed to displaying a pattern to its behavior. A more rigorous definition will be provided in the discussion on control charts.)

Populations and Samples

In statistics, a population is treated as a relatively large set of items from which a smaller sample may be extracted. A population includes all the data that might be collected concerning a particular characteristic, while a sample is a subset of this information. Examples of a population include the number of people watching the same television program, the number of headlights made for automobiles in a year's time, the measured diameters on a 1000 lot run of drill bits, or the number of grains of sand on the beach in Hawaii. If we denote x as a numerical characteristic of some population N, then the population mean μ can be expressed as

$$\mu = \Sigma \frac{x}{N}$$

where the summation is taken over the entire population x(1), x(2),. . ., x(n). The mean is sometimes called the expected value E(x). For the previous examples of discrete variables it may also be expressed using probability terms as

$$E(x) = \Sigma\, x(i)p[x(i)]$$

which is a weighted average of the possible values x(1), x(2),. . ., x(n) of the variable x. The weights are the associated probabilities of p[x(1)], p[x(2)],. . ., p[x(n)]. For the processes A and B shown in Table 1, the means are calculated as shown below:

Process A:

$$
\begin{aligned}
E(x) = {}&(2)(0.001) + (3)(0.001) + (4)(0.002) + (5)(0.003) \\
&+ (6)(0.004) + (7)(0.005) + (8)(0.007) + (9)(0.01) \\
&+ (10)(0.011) + (11)(0.015) + (12)(0.02) + (13)(0.05) \\
&+ (14)(0.01) + (15)(0.574) + (16)(0.1) + (17)(0.05) \\
&+ (18)(0.02) + (19)(0.015) + (20)(0.012)
\end{aligned}
$$

$$E(x) = 14.743 \text{ marbles/bag}$$

Process B:

$$
\begin{aligned}
E(x) = {}&(10)(0.001) + (11)(0.0002) + (12)(0.0005) + (13)(0.001) \\
&+ (14)(0.0025) + (15)(0.9814) + (16)(0.0025) \\
&+ (17)(0.001) + (18)(0.0005) + (19)(0.0002) \\
&+ (20)(0.0001)
\end{aligned}
$$

$$E(x) = 14.850 \text{ marbles/bag}$$

Both of these processes have very similar mean values for the number of marbles that are loaded into each bag. As was apparent from the earlier discussion on the distribution of the number of balls per bag for each process, however, process B is definitely superior to process A from the standpoint of uniformity and potential customer satisfaction. It was also apparent that what is needed to more completely characterize these two processes is a statistic that describes the process variability, not just the mean value of the process output.

The variance is a measure, in squared units of *x*, that describes the variability that exists in a population *N*. (For those people that find quantities such as "squared bags" or "squared temperature" distracting another statistic will be described in following sections that avoids this terminology. For the purpose of making quality-level comparisons, however, the units of the statistic should not cause excessive concern.) Using the entire population of entities in the summation gives the following equation for the variance:

$$\sigma^2 = \Sigma \frac{(x - \mu)^2}{N}$$

or

$$\sigma^2 = V(x) = E(x - \mu)^2$$

Again, using probability terms, the variance can be expressed as

$$V(x) = \Sigma [x(i) - \mu]^2 p[x(i)]^2 = \sigma^2$$

which is a weighted average of the squared deviations of $x(1), x(2), \ldots, x(n)$ from the mean μ. A useful simplification for some calculations is

$$\sigma^2 = \Sigma x(i)^2 p[x(i)]^2 - \mu^2$$

which is derived by expanding the $[x(i) - \mu]^2$ term in the preceding equation and simplifying the results. Using this method with the marbles per bag example gives the following results for the two processes.

Process A:

$$\begin{aligned}
\sigma^2 = \ & (4)(0.001) + (9)(0.001) + (16)(0.002) + (25)(0.003) \\
& + (36)(0.004) + (49)(0.005) + (64)(0.0070) \\
& + (81)(0.01) + (100)(0.011) + (121)(0.015) \\
& + (144)(0.02) + (169)(0.05) + (196)(0.1) \\
& + (225)(0.574) + (256)(0.1) + (289)(0.05) \\
& + (324)(0.02) + (361)(0.015) + (400)(0.012) \\
& - (14.743)^2
\end{aligned}$$

$$\sigma^2 = 221.511 - 217.357 = 4.155 \text{ (marbles/bag)}^2$$

Process B:

$$\begin{aligned}
\sigma^2 = \ & (100)(0.0001) + (121)(0.002) + (144)(0.0005) + (169)(0.001) \\
& + (196)(0.0025) + (225)(0.9814) + (254)(0.0025) \\
& + (289)(0.001) + (324)(0.0005) + (361)(0.0002) \\
& + (400)(0.0001) - (14.85)^2
\end{aligned}$$

$$\sigma^2 = 222.783 - 220.523 = 2.26 \text{(marbles/bag)}^2$$

Fortunately, these types of calculations are automated on many electronic calculators or computer systems so that it is rarely necessary to do more than just input the raw numbers to obtain the variance of a set of data.

The standard deviation σ is in the same units as the variable x (which may be more appealing to some people) and is the square root of the variance, or more specifically:

$$\sigma = \left[\sum \frac{(x - \mu)^2}{N} \right]^{1/2}$$

The standard deviations for process A and B are 2.038 and 1.504, respectively. These values are not that different because of the drastic effect that the 100 bags with zero marbles has on process B. If these 100 bags were eliminated through postprocess inspection or distributed proportionately about the process mean instead of occurring at the extreme end of the distribution, then the variance would be 0.03 and the standard deviation would be approximately 0.2. This type of situation demonstrates the need to rely on more than just the pure statistical parameters when evaluating a process. Even though the statistics provide numerical values that are very useful for system analysis, it is also necessary to remember what they mean and how they can be misleading at times. The integration of common sense with statistical techniques will usually avoid this type of pitfall.

A sample is a relatively small group of entities, n, that is taken from some larger population N. Because samples are usually significantly smaller than populations they are much more convenient to work with and in some instances, sample data may be all that is available. Obviously, it is easier to analyze a relatively small number of items than it is to evaluate a very large population—even if the only consideration is the data entry process. However, it is necessary to obtain a sufficient number of samples to arrive at an accurate estimation of the population's characteristics. The crucial factor to consider is the appropriateness of using a particular sample or group of samples to characterize an entire population. Selecting 3 parts out of 1000 to characterize a manufacturing process may be a reasonable approach for a very tightly controlled process that exhibits little variation with respect to the process tolerances. However, it also can be as practical as selecting a handful of sand from a beach in San Diego to describe all the beaches in California. As might be expected, the more variability that exists within a population the larger the number of entities that should be included in the sampling process to assure the reliability of the sampling procedure. (Process variability and stability are related since small variability is associated with "good" stability, but a stable process does not dictate a particular degree of variability.)

The mean for a sample is called the average, $\bar{x}$, and the calculation involves all n values of the sample.

$$\bar{x} = \sum \frac{x(i)}{n}$$

or

$$\bar{x} = \frac{1}{n} [x(1) + x(2) + x(3) + \cdots + x(n)]$$

The measure of variability of a sample is expressed in terms of the sample variance, s^2, which is again in the squared units of x.

$$s^2 = \sum \frac{(x - \bar{x})^2}{n - 1}$$

The sample standard deviation s is calculated by taking the square root of the variance, or

$$s = \left[\sum \frac{(x - \bar{x})^2}{n - 1} \right]^{1/2}$$

The numerical values that are calculated using the above equations are statistical estimates of the actual parameters of a particular population. They can be quite good methods of representing the characteristics of a process as long as they are treated carefully. Often, there is a tendency to take a small sample of data that describes the condition of a process at some point in time and do a quick average and standard deviation calculation, perhaps on a hand-held calculator. Depending on the stability of a particular manufacturing operation, this can be a misuse of statistics that will result in erroneous assumptions about process quality. Later sections of this text present techniques involving control charts that permit intelligent decisions to be made on the stability of a process at a particular point in time.

Probability Distributions

Experimental error, or process variation, is the fluctuation that occurs within a process when the environmental conditions are maintained as constant as possible. Typically these process variations are attributable to a variety of error sources such as measurements uncertainties, unavoidable changes in conditions, necessary shifts in procedures, equipment wear, and so on. Careless mistakes or other people problems are not included in this category for the purposes of this discussion. Examining these normal variations or process errors in a manufacturing operation provides valuable insight into the reasons behind the lack of perfection in the process output. In addition, it provides an important technique for evaluating the results of attempts at improving the quality of an operation, as well as offering a set of performance indices that can be used to detect trends and predict problems before they become overwhelming. ("Assignable cause" variation is a term used to describe the "unusual" variations occurring within a process that are attributable to a "nonnormal" condition.)

Probability distributions provide a convenient way of graphically depicting the characteristics of a particular process. The shapes of the discrete probability distributions shown earlier in Figure 1 are significantly different, and they can be used to estimate the probability that the quality of a process will fall within a specific range. In the marble example given earlier, it can be seen that process A is reasonably well centered about the desired value (15 marbles), but there is a significant probability that a lot of the bags will have too few marbles:

$$p(1) + p(2) + \cdots + p(15) = \frac{2,290}{10,000} = 22.9\%$$

Similarly, 19.7% of the bags will have too many marbles, which will not disappoint the customers but will increase manufacturing expenses. Process B is a much more tightly controlled operation in that only 1.86% of the bags will have an incorrect number of marbles. In addition, approximately half of these bags will have zero marbles, probably due to a bag-handling malfunction, which may not be a problem since this can be an easy condition to detect on inspection.

Figure 2.3 shows a continuous probability distribution that is quite useful called the normal distribution. This distribution is defined by the equation

$$p(x) = Ce^{-[(x - \mu)^2/2\sigma^2]}$$

where C is a constant and μ and σ^2 are the population mean and variance, respectively. This entity is also called the Gaussian distribution and plays an important role in the theory of statistics. It is a symmetric curve with its highest ordinate at the point where the abscissa equals μ and it tails off to zero on both sides. This distribution has been studied thoroughly and is important in statistics because it describes a random cause system that is typical of many manufacturing processes. In fact, there is a tendency for many real-life distributions to be very similar to a normal distribution. Also, many commonly used statistical procedures are insensitive to deviations from the theoretical normal distribution. Therefore, only approximate normality is required for these techniques to be applicable. Finally, knowing that a particular process can be accurately characterized by the normal distribution can help to avoid the requirement for a large amount of historical reference information.

Manufacturing process variations typically arise from a large number of error sources of varying degrees of significance. In addition, processes that are sensitive to a particular group of error sources, or disturbances, at one point in an operation may be insensitive to these same factors at another part of the manufacturing cycle. The continuous interaction of these errors of varying significance results in a distribution of the process characteristics that approximates a normal distribution. This is due to a mathematical phenomenon called the central limit effect. This mathematical theorem is based on the assumption that the total process error is a combination of a large number of individual errors (which is true for many manufacturing operations). If each of these individual error sources contributes only a relatively small percentage to the total error, then it is possible to approximate the overall error for a specific set of operating conditions as a linear function of independently distributed component errors:

$$e = c(1)\, e(1) + c(2)\, e(2) + \cdots + c(n)\, e(n)$$

where the c's are constants. Under these conditions, the central limit theorem states that the distribution of this linear function of errors will approach normality nearly independently of the individual distributions of the individual error

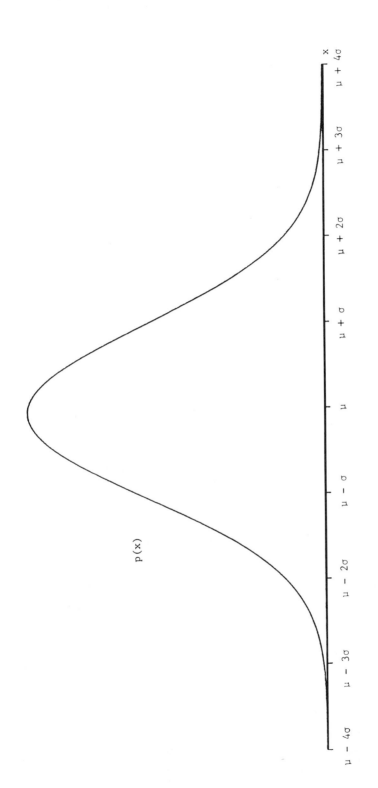

Figure 2.3 Graph of the normal distribution.

components. The attractive feature of this phenomenon is that even though the individual error sources may not be characterized by a normal distribution, the combined result of these disturbances is approximated by the normal distribution.

As shown by the equation used above to describe the normal distribution, once the mean μ and the variance s^2 of a normal distribution are known, the entire distribution is defined. Because of this situation, the notation $N(\mu, \sigma^2)$ is used in statistics as a symbolic way to describe a normal distribution which has a mean of μ, and a variance of σ^2. Figure 2.4 shows a variety of normal distributions which have different "sizes." However, all of the curves are characterized by the same basic "shape" because they are all generated by the same basic equation with only a change in the value of the constants. For each of these curves the standard deviation σ measures the distance from the mean μ to the point on the curve at which the slope changes from positive to negative (the point of inflection).

A significant benefit that arises from all this concern about having a normally distributed manufacturing operation is the ability to make accurate predictions about the future operation of the process. This occurs because it is relatively simple to determine the probability that an event will occur in the process which deviates from the process mean by a certain amount in a given direction. For instance, the probability of a process characteristic deviating from its mean value by an amount equal to one standard deviation, or one sigma, is found by measuring the appropriate area under the tail of the curve. For a one sigma deviation the area is measured beginning with the point on the abscissa equal to the mean plus one standard deviation. This is equal to 0.1587, or approximately one-sixth of the time. For a two-sided probability, that is, the situation in which an occurrence exists outside the range bounded by the mean plus or minus one standard deviation, the area under both tails of the curve is summed. This can be accomplished by doubling the area under one end of the curve because the normal distribution curve is symmetric about the mean. This is equal to $2 \times 0.1587 = 0.3174$, or about one-third of the time. Similarly, the probability of a deviation occurring at a point equal to the mean plus two stand deviations is equal to 0.02275 for a single-sided measurement or roughly one time in twenty.

While these relationships are of theoretical interest, it is not practical to expect someone to be very enthusiastic about spending a lot of time measuring the areas under curves for different deviation ranges. Fortunately, standard tables are available in most statistics texts and math handbooks to permit the user to look up the desired information rather than perform the otherwise necessary calculation. However, it is also impractical to generate a table for all the possible combinations of means and variances that might be of interest. To avoid this problem, a process called normalization is utilized to produce a standard normal distribution that can be applied in a wide variety of cases. While the normal distribution is centered about the process mean, the standard normal distribution is centered about zero. In addition, the scale of the abscissa is changed from units

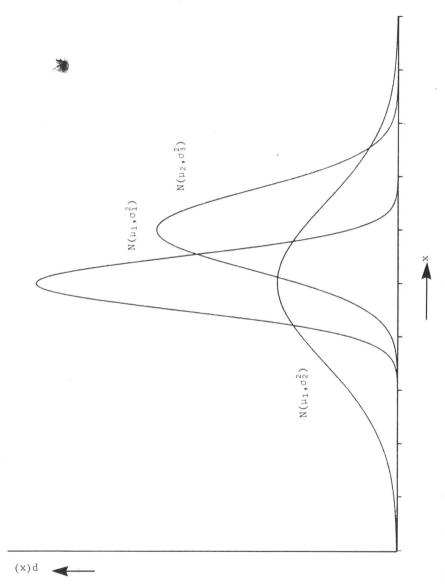

Figure 2.4 Normal distributions with different means and variances.

of sigma to the simple numerical values 1, 2, 3, etc., as shown in Figure 2.5. This is accomplished through a shift in the variable used to plot the normal distribution curve. This is accomplished by creating a variable z that is defined to be equal to $(x - \mu)/\sigma$. This means that the distribution of z is $N(0,1)$ because of the mathematical transform achieved through the subtraction of the process mean and the division by the standard deviation. In the same manner, the probability that an occurrence is greater in value than the mean plus one standard deviation becomes

$$p(x > \mu + \sigma) = p([x - \mu] > \sigma) = p(\{[x - \mu]/\sigma\} > 1) = p(z > 1)$$

which was shown previously to be equal to 0.1587. The practical value of all of this is that once the mean and variance are known for a normally distributed process, then it is relatively easy to determine the probability of the occurrence of a given event, for example, the probability that the process will produce an out-of-tolerance part.

Table 2.2 shows the single-sided tail area of the standard normal distribution for a limited range of values. For example, consider a situation in which a normally distributed process is used to fabricate valves for automobile engines. Through historical measurements, the mean diameter of the population of valve stems produced by the process has been calculated to be 0.300 in. while the standard deviation of the operation is 0.001 in. Therefore, the probability that a valve with a diameter greater than 0.3015 was produced as a part of the given population, or will be produced in the future (assuming a constant process) may be determined by calculating z as shown below:

$$z = \frac{0.3015 - 0.300}{0.001} = 1.5$$

From Table 2.2 it is seen that the area value corresponding to a z of 1.5 is 0.0668 so the probability of having a valve stem with a diameter greater than 0.3015 is 6.68%. In a similar fashion, the probability of having a valve stem with a diameter that is either 0.0015 in. larger or smaller than the target value of 0.300 in. is 13.36%.

In the previous example it was assumed that the standard deviation of the population was known. In other instances, it may be that only the standard deviation of a sample of the population is available. In this case the Student's t distribution, which is shown in an abbreviated form in Table 2.3, may be used. Using s, the standard deviation of a sample of measurements, the value for t is calculated as shown below:

$$t = \frac{x - \mu}{s}$$

If a sample of 10 measurements was used to generate the standard deviation from the previous example (s = 0.001) and the mean of the population is known

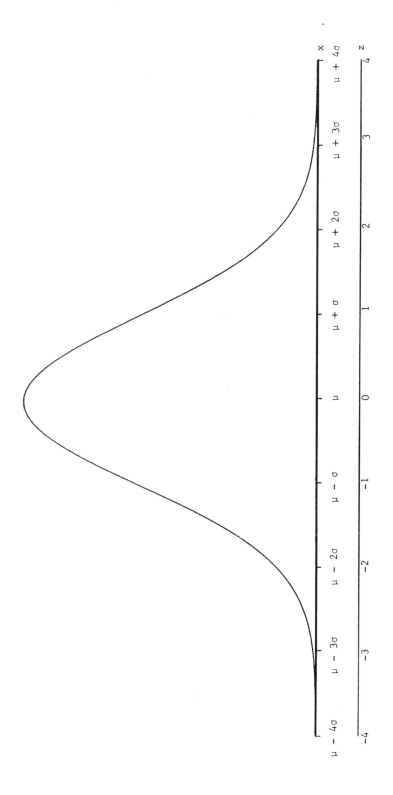

Figure 2.5 Standard normal distribution.

Table 2.2 Tail Area of
Standard Normal
Distribution

z	Area
0.0	0.5000
0.5	0.3085
1.0	0.1587
1.5	0.0668
2.0	0.0228
2.5	0.0062
3.0	0.0013

Source: Box, Hunter, and Hunter
(1978).

to be 0.300 from historical data, then the problem can be reworked based on
the current sample of measurements, where

$$t = \frac{0.3015 - 0.300}{0.001} = 1.5$$

Table 2.3 is used by finding the row equal to one less than the number of samples
in the data set (degrees of freedom) and moving over to find an entry in the body
of the table equal to the t value calculated above. Moving vertically to the tail
area probability row gives the probability of the occurrence of the particular event.

Table 2.3 Probability Points of the t
Distribution with v Degrees of Freedom

v	\multicolumn{4}{c}{Tail Area Probability}
	0.4	0.25	0.1	0.05
1	0.325	1.000	3.078	6.314
3	0.277	0.765	1.638	2.353
5	0.267	0.727	1.476	2.015
7	0.263	0.711	1.415	1.895
9	0.261	0.703	1.383	1.833
11	0.260	0.697	1.363	1.796

Source: Box, Hunter, and Hunter (1978).

If an exact match is not found, then interpolation is used to calculate the correct value. In this example the t value of 1.5 gives a probability value that lies between 0.05 and 0.1. Interpolating gives a closer approximation of about 8.7%. As might be expected, this probability value is larger than the one calculated previously because of the use of the standard deviation of the sample rather than the standard deviation of the population. In this case, it was assumed that the population mean was available from historical data. This assumption implies that it is expected that the process mean normally remains constant, although the scatter in the process may vary. If this is not true, it is necessary to rely upon the available data to make the best estimate possible and to use caution until sufficient supporting data is available.

External Reference Distribution

The previous examples have focused on using probability distributions to describe the condition of a process at a given point in time. The logical extension of this activity is to utilize this technique to compare the quality of an operation under different sets of circumstances. In order to evaluate the results of a system modification it is necessary to have a baseline condition that can be used for the judgment process. The external reference distribution is the probability distribution of the system being analyzed when operating under normal conditions. This information provides a relevant reference set that can be used to determine if statistically significant changes have occurred in the operation of a process. The intent is to obtain a method of deciding if the differences in the process characteristics due to a change in the operating conditions are larger than what would usually be expected given the previous history of the process.

Assume that process A has been modified in some fashion and a test has been run to fill 100 bags with marbles. The number of marbles in each of these bags was tested and the question that arises is whether or not the new process is significantly better than the old process. As shown above it is possible to calculate values for the mean and variance of this sample. Assume that the mean value is still reasonably close to 15 marbles per bag as was previously observed, but the variance of the sample is now $3.44(\text{marbles/bag})^2$. Using the historical data it is now possible to decide whether the change in the variance is statistically significant or not. If the original data is broken up into groups of 100 runs, it is possible to calculate the variance for each of these data sets and determine how often the variance would have been 3.44 or less when sampling the process under the old operating conditions. If this condition occurred relatively infrequently, then it can be safely assumed that the changes in the process are statistically significant. On the other hand, if this variance value occurs rather frequently, then no statistically significant alteration was caused by the process modification. Those process changes that cause an area of operation between these two extremes are not particularly interesting because, like the statistically insignificant changes, they only provide information on what does not help the process quality. The advantage to this technique is that it does not require random sampling for validity. The only assumption is that the process' operating conditions are unchanged throughout the testing (with the exception of the alteration used to evaluate the modified system). The disadvantage to this approach is that an extensive set of relevant prior data is required for the system being examined.

Summary

An interest in monitoring or controlling the quality of a process is uniform throughout the manufacturing community. Unfortunately, one of the difficulties in improving the quality of a process is figuring out where to start and how to decide whether the implemented changes are beneficial. This chapter has presented statistical techniques that can be employed to describe the condition of a process and that may, if applied appropriately, be utilized to evaluate the results associated with process modifications. However, this chapter is only an introduction to the application of statistics in manufacturing, and other techniques will be presented later that build on this material. In addition, other approaches will be offered that do not require an extensive amount of historical data at the expense of additional assumptions about the manufacturing process.

In general, one of the primary goals that is inherent in the effort to improve the quality of a manufacturing operation is the reduction of process variation. If this characteristic can be minimized, other adjustments can usually be employed that will shift the mean of the operation to the desired operating point. In addition, once a condition of stability is achieved, techniques such as deterministic manufacturing can be employed to reduce fabrication costs and improve the overall quality of the manufactured products.

Also, the statistical characterization of process distributions can be applied to incoming products as a means of evaluating the quality of a vendor's process. Assuming a vendor's manufacturing process is operating in a stable fashion, it is possible to compare process characteristics at different points in time as a means of avoiding 100% inspection of the incoming products. Since most manufacturing processes can be accurately characterized by the normal distribution, deviations from this condition may be cause for further investigation. A nonuniform distribution of products is probably indicative of a "sort and sift" operation, in which the usual process output isn't always within a needed tolerance, so postprocess inspection is substituted as a means of delivering acceptable units. While this does not cause problems with the use of the delivered products, it does indicate that the user's tolerances are tighter than what the vendor can routinely deliver. This probably means that the buyer is paying for additional inspection operations and that the ability of the supplier to deliver products with the required quality attributes at some time in the future may be questionable.

CHAPTER 3

Deterministic Manufacturing

Introduction

To the layman, manufacturing is frequently thought of as being some rather obscure operation that occurs in large buildings called factories. It is recognized that raw materials are delivered from some unknown location and introduced into the facility where they are mysteriously converted into suitable products such as automobiles, stoves, televisions and so on. While the method through which this magical transformation takes place generally is not understood, it is assumed to occur through the use of devices that are generically labeled machinery. In addition, recent publicity in the news media gives the impression that many factories are no longer productive and that the country has lost its previously held position of leadership in the competitive worldwide manufacturing market.

While the relative status of the different members of the worldwide manufacturing community may be debatable, it is true that the intent in most manufacturing facilities is to cause machines (which may appear to have minds of their own) to function in a systematic manner. It is assumed that this will permit the factory operation to be optimized for a given resource investment and therefore generate a profit. For the purposes of this text, the definition of manufacturing shall be limited to the process of using machines to make a variety of piece parts, such as door knobs, aircraft engines, or the components of another machine such as the leadscrew shown in Figure 3.1. In this instance, the focus is centered on

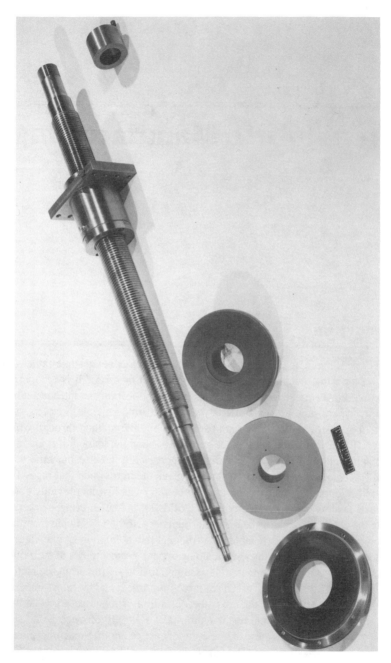

Figure 3.1   Parts of a machine leadscrew (courtesy of Martin Marietta Energy Systems, Inc.).

piece-part manufacturing operations in order to limit the scope of the book to a manageable size. The decision not to include material on the process industries, such as petroleum refining, was made for the purposes of simplicity and does not imply a lack of importance associated with these operations. In addition, the philosophical concepts involved are equally valid in both types of environments.

The underlying idea encompassed by the deterministic manufacturing philosophy is that events occur in the manufacturing world because of specific causes, as opposed to random chance, even if the individual observers involved do not recognize the antecedent factors. This approach does not conflict with the random occurrence that is discussed by statisticians. The statistician is well aware that events occur because of "stimuli." However, he is frequently unable to predict with absolute certainty when a given stimulus will occur, which leads to the "random event" terminology. The concept of determinism is similar to the "action and reaction" theory taught in physics courses. Also, it is an idea that most manufacturing personnel will accept without a second thought as being almost an understatement of the obvious. In fact, most personnel involved in manufacturing have said something like "we've got to find out what's causing that problem" at some time or another, because it is readily recognized that malfunctions do not occur without a reason. These cause and effect laws of nature or deterministic philosophies are widely embraced in industry because it is a matter of common sense. Unfortunately, while the underlying idea may be obvious to everyone and the concept is intuitively understood, it is not utilized to its full potential. The usual approach is primarily to deal with the situation after the fact and along the lines of "let's find out what caused the system to malfunction" rather than using the information ahead of time. We would all like to know what caused a machine to make a defective part rather than the intended product, but the classical manufacturing approach is to wait for the postprocess inspection information to characterize the process before attempting to detect whether a problem exists or not.

Obviously, the ability to predict accurately the shift in an important process variable is a useful tool in a manufacturing operation if an excursion in this parameter can cause an expensive disruption in the system's performance. The ability to avoid potentially degraded performance is attractive because no one likes waiting for something serious to go wrong before initiating remedial actions. Closed-loop control systems are widely used to control parameters such as motor speed and current, fluid temperature, table position, etc., in many manufacturing processes. However, these are parameters that are widely understood and the control function is defined more by the desired operation of the machine than it is by the manufacturing process. The result is that a machine may have a tool path accuracy of 0.0002 in. but the process has difficulty in producing a product to an accuracy of 0.002 in. It is unusual to encounter a situation in which the key process parameters are being monitored and used to characterize and control the manufacturing operations through adjustments to the individual machines.

While process adjustments are frequently employed in an attempt to "center a process," this is usually done in response to postprocess inspection information rather than an in-process measurement of the critical process parameters.

For example, numerically controlled machine tools utilize position and velocity feedback transducers to control the relative locations of the machine's slides and to a lesser extent the relative locations of the workpiece and the cutting tool. However, it is unusual to encounter an application in which tool wear is monitored in real time and used to control the metal-cutting process. Instead, the conventional approach is to estimate the quality of the cutting tool through the measurement of a control dimension on the workpiece after the part is completed. This postprocess inspection data is then used to determine adjustments that are incorporated into the process in an attempt to obtain/maintain the desired part dimensions on subsequent parts. This technique can be performed in a statistically sound fashion and it is a valid approach as long as the processing conditions remain relatively constant from one part to the next. Unfortunately, another possibility is that this processing approach can produce rejected workpieces if an unexpected shift occurs in a key process variable which is not being monitored. Part of the difficulty is that the effect (excess stock due to tool wear) of different process variables is being monitored rather than the actual values of the process variables. For example, a shift in workpiece hardness from one part to the next will result in an unexpected amount of tool wear and excessive material being left on a part. This condition would not be detected until it was too late unless in-process measurements were performed prior to removing the part from the machine. (In some situations it is possible to reload a part in a machine and attempt to fix it but relying on a rework cycle is not an effective method of manufacturing quality/economical products.)

In contrast, the deterministic manufacturing approach to product fabrication involves the utilization of process monitoring to estimate, in real time, when a significant alteration has occurred within some portion of the manufacturing cycle. The intent is to use limited in-process inspection coupled with the monitoring of the key process parameters as a substitute for extensive postprocess inspection. (The simple substitution of extensive in-process inspection activities for extensive postprocessing inspection operations is a step in the wrong direction.) The advantage to this type of operating procedure is that it is not necessary to produce a certain amount of scrap before the system is able to detect that a problem exists. Instead, process shifts are detected as they occur and the potential effect on the process output is estimated without having to complete the fabrication of a workpiece. At this point adjustments can be made in the processing operations to accommodate the changed parameters or else maintenance can be performed to remedy the cause of the problem. Another benefit to this mode of operation is that a sampling procedure can be implemented to inspect a limited number of finished parts while depending on the monitoring process to provide the necessary assurance that the system is still operating the way it should.

A good example of the simplicity with which this technique can be applied is shown by a torque monitoring system that was installed by the Spicer Heavy Axle Assembly Division of the Dana Corporation. [1] This 250,000 square foot facility assembles over 300 light and heavy truck axles per day. A basic requirement of this operation is the accurate torquing of 15,000 to 17,000 bolts per day. Previously, this plant utilized periodic manual checking of bolt torques to establish process quality. However, whenever an anomaly was detected it was necessary to hand check all the assemblies that had been fabricated since the last monitoring cycle. As might be expected, this departure from normal operations introduced a significant perturbation in the production rate. To avoid these difficulties, a torque monitoring system was installed that automatically records the torque employed with every bolt. In addition, indicators are provided that permit the system operators to recognize shifts in the process so that adjustments can be made before significant errors are produced. The results of this operating method have been a 20% increase in productivity and total quality control, as well as a permanent record of the key parameters for every axle that is produced.

While the axle assembly operation does provide a clear example of the benefits that can be obtained in a relatively simple manufacturing system, significant advantages are also available to more complex operations as will be shown later in this chapter. The only difficulty in applying this procedure to the more complex manufacturing tasks is that it may not be a simple problem to identify and monitor the necessary key process parameters. Without this parameter information it is difficult to construct an accurate model that can be used to predict the system output. However, if these parameters cannot be established and monitored, then the prospects of improving the quality of the operation using any technique are not very high, because it is difficult to improve upon a system that is not understood. As Bill Rasnick says, "Know thy process."

Types of Manufacturing Operations

As mentioned previously, this text is primarily concerned with manufacturing operations that are used to fabricate piece parts. However, this scope is still quite large and may be a distraction unless it is stressed that the exact machine configuration and/or task that it is designed to perform is less important than the basic concept presented. An analogous situation exists with numerical control systems for machine tools. The same basic control system can be used to control a slant-bed lathe, a milling machine or a machining center. The basic requirement is that the system perform the necessary interpolation task that is required to convert the part program commands into specific axis commands and to provide a drive signal for the machine servo systems. In addition, the control must respond correctly to the manual commands from the operator as well as deal with the specific requirements of a given machine tool. However, the basic control strategy does not change from one machine to the next. Only the details of implementation are altered to meet the unique needs of a particular situation.

Three examples of metal-working machines that may be used to produce a variety of piece parts are shown in Figures 3.2-3.4. The specific type of operation that is performed by each of these machines is relatively unimportant. The more important concept is that many machines designed to accomplish one job have much in common with other machines in dissimilar industries. If the basic underlying method of applying the deterministic manufacturing approach is understood with respect on one type of system, then this operating philosophy can be readily transferred to other types of operations. For instance, the machine represented in Figure 3.2 is a vertical machining center that positions the workpiece beneath a tool head that travels in the vertical direction. This is similar to the method of operation of a sewing machine or a punch press, although the rotary motion of the tool associated with the machining center adds extra

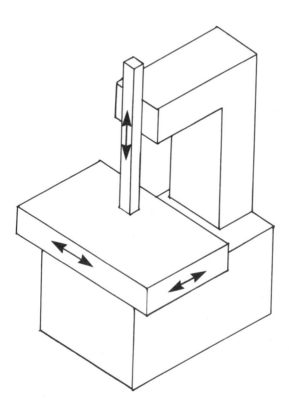

Figure 3.2 Vertical machining center.

complexity to the mechanism. In the same manner, Figure 3.3 shows a high precision inspection machine that has been converted into a lathe for precision diamond machining operations. Not only are the axes motions the same for the inspection and diamond turning machines but, in this case, the hardware is identical even though the functions may appear to be quite different at first glance. Similarly, the robot-loaded lathe shown in Figure 3.4 shares common attributes with a robot-controlled painting station that is used on an automotive assembly line.

The common theme throughout these and many other operations, is that the crux of the task is the positioning of a workpiece and a tool of some type at specific locations with respect to each other. Then, depending on the specific function of the particular machine, work is performed on the part to transform it into the desired condition. Whether the manufacturing requirement is to saw the end off of a piece of lumber perpendicular to its sides or to grind a fuel injector nozzle to a size tolerance of 0001 in., the machine performance is a critical portion of the manufacturing cycle. It should be apparent that the relative accuracy with which the workpiece is positioned and the machine's motions are controlled is more important than the particular type of manufacturing operation that is being discussed. This does not mean that machine tool path is the primary cause of manufacturing errors. While tool path can be an important consideration, other factors such as fixturing errors can overshadow the problems caused by the machine's positioning inaccuracies. The key objective of most manufacturing operations is to correctly position the workpiece and the tool in relationship to each other during the fabrication cycle so that the desired configuration is obtained on the finished part in spite of the existence of process disturbances. While various types of operations do require significantly different levels of absolute accuracy, and the degree of difficulty in achieving these accuracy levels varies with each particular task, it is possible at times to lose an important perspective. The main concept to keep in mind is that accuracy in relation to tolerance should be the major focus and this is independent of whether the job is to saw 2 × 4s or to machine reflective metal optics with a single crystal diamond tool.

Machine Accuracy

As mentioned above, machine accuracy is one of the important contributors to process quality. Machine accuracy primarily relates to the ability of the machine to position its moving members at specific points in space in a static and dynamic fashion. A generic term that is used to describe this condition is tool path accuracy. In general, this characteristic is influenced by two basic categories of errors. Quasi-static errors are process disturbances that change relatively slowly and are therefore said to have long time constants. The result of this type of system error is usually observed as the degradation of part form due to the inaccurate positioning of the tool with respect to the workpiece. This class of errors is related

Figure 3.3 Precision diamond turning machine (courtesy of Martin Marietta Energy Systems, Inc.).

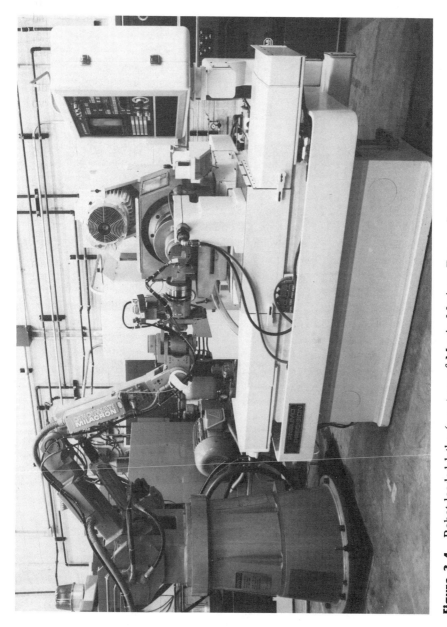

Figure 3.4 Robot-loaded lathe (courtesy of Martin Marietta Energy Systems, Inc.).

to the machine's structure and design, and the results of these imperfections are often manifested in unexpected ways. These errors are due to the geometry and kinematics of the machine, the slowly varying forces that act on the machine, and thermally induced strains in the machine tool. The dynamic errors in a manufacturing system have relatively short time constants. This means that they occur at a relatively high frequency, and include items such as vibration, spindle errors and axes motion errors related to the performance of a machine's servo system.

The geometry errors associated with the travel of a machine's moving elements can be discussed in relation to the typical machine axes shown in Figures 3.5 and 3.6. A linear slide, such as the one represented in Figure 3.5, typicaly utilizes a set of guideways that provide support for the axis load as well as directional control. (The pair of square guides shown in Figure 3.4 are intended to represent the slide guide mechanism and do not represent the configuration used on an actual machine. Although some machines are commercially available that utilize double-vee guideways, most machine slides use a configuration that incorporates a design such as a flat and a vee to avoid overconstraining the system.) With the type of linear carriage shown in Figure 3.5, it is possible to measure six individual error elements due to the six degrees of freedom present (three degrees of freedom for translation, and three degrees of freedom for rotation). The angular error motions are described as roll, pitch and yaw, as shown by Figure 3.5, and are usually relatively small for a good quality linear slide. An error motion that is measured along the axis of travel is commonly called the positioning error, while the errors perpendicular to the axis of motion are discussed as straightness errors in the direction of the corresponding axis. For instance, an x-axis

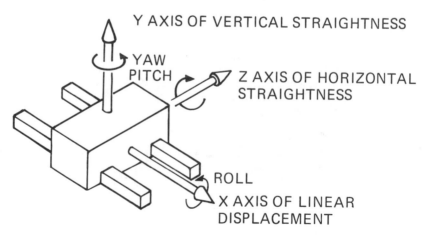

Figure 3.5 Linear axis error motions (courtesy of National Bureau of Standards).

slide on a three-axes machine would be said to have positioning errors in the direction of the axis travel and straightness errors in the y and z directions.

Figure 3.6 shows a similar notation for a rotary axis, such as a spindle or rotary table, in which there are still six degrees of freedom. In this case, the linear errors correspond to a shift of the center of rotation of the axis center as the axis rotates. The error in the z directon results in a shift in the plane of rotation but does not degrade the accuracy of rotation. Errors in the x or y directions mean that a point located at an arbitrary point on the axis would not generate a perfect circle as the axis rotates, which would result in roundness errors on a workpiece. In comparison with a linear axis, a good quality rotary axis has translation errors that are relatively small, but the rotational errors are more significant. The rotational error in the plane of rotation corresponds to the positioning error associated with a linear axis, while the other two angular errors result in a tilt or wobble motion about the respective axes.

Static deformations occur in all machines to varying degrees. These unwanted displacements occur because of a change in the machine loading, as a result of the shifting positions of the machine components, or because of the loading or unloading of parts and fixtures. As the machine members move, the change in

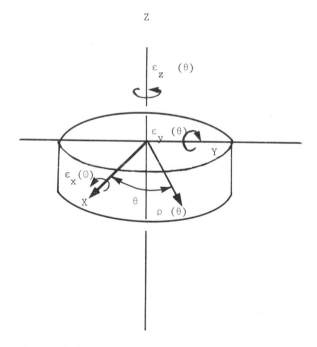

Figure 3.6 Rotary axis error motions (courtesy of National Bureau of Standards).

position shifts the force distribution on the machine. The resulting stress variation causes machine deflection and bending. A similar reaction occurs when a
sufficiently heavy part or other component is added or removed from the machine.
Clamping mechanisms used to secure parts, fixtures, or other machine elements
can also cause machine deformations.

The temperature environment has long been recognized as a key parameter for
dimensional measurement [2]. However, it is only relatively recently that the temperature effects on machine tools have been given serious consideration outside
of specialized areas such as precision diamond machining and ruling engines.
The thermal gradients in a machine also can cause problems in a manner similar
to the varying force distribution situation. The gradual expansion or contraction
of the machine is a problem in itself due to the shifting of position references.
However, a nonuniform temperature distribution causes additional stability and
repeatability problems due to the generation of internal stresses. The result is
often bending and twisting of the machine members that cause insidious machine
geometry errors because the deflections are often unrecognized by the machine's
position transducers.

The important dynamic machine errors are those related to the high-speed components of the system, such as the spindle or a rapid indexing mechanism. Other
error sources are associated with the coordination of axes motions and vibration
errors. Error motions related to work spindles tend to cause errors in the shape
or roundness of parts as the rotational motion of this axis deviates from a perfect
circle. Cutter spindles are less susceptible to this variable since the effective "high
point" produced by the cutter rotation determines the depth of cut. However,
deflection and chatter of spindle-mounted cutters can be a problem. Axes coordination difficulties result in form errors that are frequently related to axes feed
rates and are generally independent of spindle condition. Vibration is usually a
major factor in surface finish problems although it can also influence the workpiece form if the vibration causes a secondary malfunction such as the break-
down of the cutting tool edge.

Tooling Accuracy

Other factors affecting the quality of the finished workpiece are errors associated
with workpiece fixturing and the cutting tool. While the clamping force used to
secure the workpiece for machining must be sufficient to maintain the part's position on the machine, this force must not cause unacceptable distortions in the
finished product or in the machine tool. Also, the part must be clamped in a condition that is as near as possible to being undeformed ("free state") [3,4], assuming that a lack of deformation is an important characteristic of the finished product.
Any elastic deformations introduced by the fixturing process will cause the part
to return to the unconstrained shape when the fixturing force is removed, which

may cause errors in the workpiece dimensions. An important parameter in this respect is the fit between the part and the mating surface used to locate the workpiece in the proper orientation. This type of error condition can occur in essentially two basic sets of circumstances. One instance is demonstrated when a flat part is clamped to an "out-of-flat" surface. An example of the other situation is when an attempt is made to force a "nonflat" part to conform to a flat surface. In either case, the end result is the same. The stresses induced by forcing the dissimilar surfaces to mate will result in part deformation after the clamping mechanism is released and the stresses are relaxed. Another difficulty with some clamping mechanisms is that they result in local deformations of the workpiece in the area around the part–clamp interface. The effect of this process disturbance is also the generation of internal stresses in the workpiece, as well as the potential for problems in part characteristics such as parallelism, concentricity, wall thickness variation and so on.

In a similar fashion, the condition and relative location of the metal-working tool is very important. On numerical control machines it is necessary to position the machine slides at the proper location before executing the controlling part program. This can be accomplished through a variety of techniques called tool set operations, which are implemented prior to the utilization of the tool path used to produce a particular workpiece shape with a grinding wheel or a single point tool. The inaccuracy inherent in this process produces a variety of workpiece errors that are dependent upon the particular operation. In addition, this category of tool positioning error leads to workpiece errors that are independent of the quality of the machine tool path.

Errors in the shape of the cutting tool also produce form errors in the finished part. Figure 3.7 shows a tool path that can be used to produce a concave or

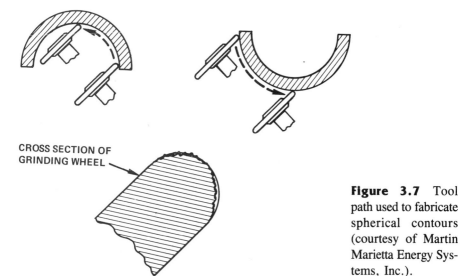

CROSS SECTION OF
GRINDING WHEEL

Figure 3.7 Tool path used to fabricate spherical contours (courtesy of Martin Marietta Energy Systems, Inc.).

convex spherical contour using a grinding wheel (or single point tool). In this case, the part is machined without maintaining a constant contact point between the part and grinding wheel as the wheel progresses around the workpiece. Figure 3.8 shows a contour grinding wheel and two graphite clips that have been used to characterize the quality of the wheel size and shape. Figure 3.9 shows an enlargement of the graphite clip that was used to characterize a worn grinding wheel. It is readily recognized that the grinding wheel cross-section is closer to a rectangular shape than the desired circular shape. Figures 3.10 and 3.11 show how these tool size and shape errors affect the quality of a spherical workpiece.

Error Sources

The meaningful error sources for a given process consist of those variables that can produce a significant deviation in one or more of the required characteristics of the end product. Excursions of these parameters, from their normal desired conditions, that do not cause significant perturbations with respect to workpiece tolerances are not considered to be important. For example, a 0.01 in. tool height error on a lathe used to machine 6 in. diameter bar stock only produces an error in the depth of cut equal to 17 millionths of an inch, which is insignificant in most circumstances. However, this same tool height error associated with a facing cut on a flat disc can leave a significant excess of stock at the center of the part (unless there happens to be a hole in the center of the workpiece).

As discussed previously, the significant process parameters for a particular manufacturing operation may include factors such as machine–machine component geometry and accuracy, temperature, vibration, and the quality characteristics of the raw material used as feedstock for the process. The goal of the deterministic manufacturing approach of product fabrication is to monitor the significant error sources and generate the appropriate action when a serious excursion occurs. Unfortunately, the difficult task, in many instances, is to identify which of a large number of potential error sources are the significant ones and which are only of secondary importance. A further complication is that the various error sources may not all be important at the same point in the manufacturing cycle. Excessive tool wear may not be meaningful just prior to a scheduled tool change, but if it occurs during a finish machining cycle it may be serious. Similarly, angular slide motion errors or straightness errors may be unimportant in some applications but they can be disastrous in other instances. This can be due to both the tolerances involved and the sensitivity of the operation to a particular error source.

To determine the most important process parameters, an estimate is made based on engineering judgment and/or historical processing knowledge and statistical tests are conducted to evaluate the validity of the initial model. Then, this system model is iteratively fine tuned through a process of model modifications and

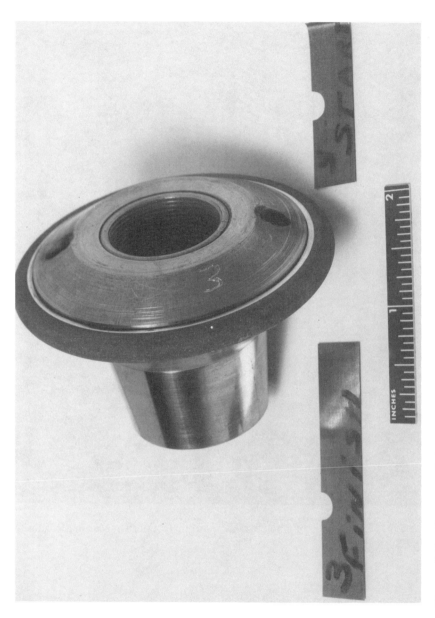

Figure 3.8 Contour grinding wheel (courtesy of Martin Marietta Energy Systems, Inc.).

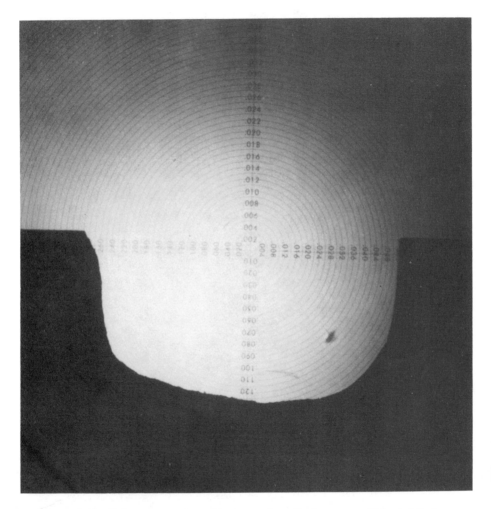

Figure 3.9   Enlargement of graphite inspection clip (courtesy of Martin Marietta Energy Systems, Inc.).

controlled testing of the process until (hopefully) a relatively few critical parameters are identified that describe the manufacturing operation to the necessary degree of precision. The model accuracy must be significantly better than the required tolerances because the model is also a process error source.

This system-testing exercise frequently provides previously unrecognized information about a particular manufacturing operation. In addition, this new information also may enable some process improvements to be implemented

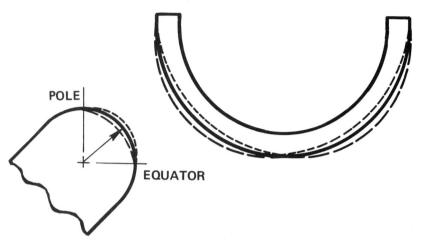

Figure 3.10   Effect of tool-size errors on spherical workpiece (courtesy of Martin Marietta Energy Systems, Inc.).

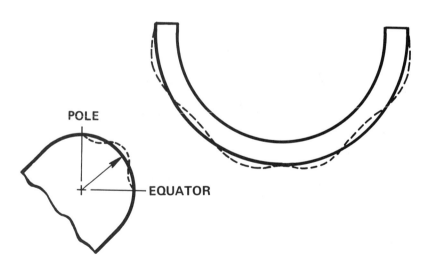

Figure 3.11   Effect of tool-shape errors on spherical workpiece (courtesy of Martin Marietta Energy Systems, Inc.).

immediately which further reduces the number of critical process parameters that require monitoring. An awkward situation that can occur is that it may be difficult to monitor some error sources directly while an operation is in progress. The previously mentioned wear on the edge of a cutting tool is an example of this type of problem. The solution to this difficulty is to monitor directly a parameter that is indicative of the state of a particular error factor and follow up this measurement with a postprocess direct measurement of the entity of interest whenever it is possible and practical.

In addition to monitoring error sources, it is frequently advantageous to monitor the system status indicators that are recognized by the system controls. This information can be valuable in trying to reconstruct the machine environment that existed prior to a malfunction. Examples of these types of parameters include servo system following errors, feeds, speeds, manual system commands, and so on. Another benefit occurs when trying to evaluate system status at different points in a manufacturing cycle, perhaps as a means of discovering the choke points in a fabrication system. Examples of these types of parameters include time spent performing a certain task or time spent waiting on something else to happen (idle time).

Process Parameters

Process parameters are descriptive entities which characterize the condition of a process. While these entities are often defined through a process of statistical calculations, it is desirable that they be directly related to some measurable process characteristic and not the result of some complex mathematical relationship that involves multiple parameters. Usually, these qualities are related to some "measure-of-goodness" or a feature with a specified tolerance for a particular manufacturing operation. (As mentioned above, it is of little use to characterize factors which do not relate to the quality of the product being produced.)

Examples of typical process parameters include cycle number, temperature, elapsed time, feed material condition, tool sharpness, servo system constants, pressure, "time of day," operator identification, control features on the workpiece, or a statistical calculation based on these or similar items. Frequently encountered statistical calculations used to describe the parameters of a process include the mean, range, maximum, minimum, and standard deviation. (These statistical parameters have been discussed in Chapter 2. In addition, other information on their utilization is included in Chapter 4.)

As an example, consider the fabrication of a graphite air bearing which is typically used with devices such as machine spindles, rotary tables and gaging probes. For the air bearing to function correctly, and maintain the necessary air film between the mechanical components at all times, it must have the proper restriction of air flow through the bearing material. Graphite is a popular material that is used for the bearing material because it is easy to machine and lap to size and it is also naturally forgiving if air pressure is lost while the components are in motion. However, either the graphite must possess the required air restriction characteristics or additional external restriction must be provided. Since external restriction devices are expensive, a typical approach is to impregnate the graphite with a substance such as pitch [5]. This material soaks into the graphite and provides a uniform permeability or resistance to penetration of the pressurized fluid. The question that arises during the impregnation process is "when to stop." Obviously, it is undesirable to plug up all of the aperatures in the graphite, but it is also necessary to provide the correct amount of restriction to the air flow. In this instance, the key process parameter is an entity that is difficult to measure directly. Fortunately, the effect of the graphite permeability can be measured with a flow meter that monitors the flow through a localized portion of the graphite. In addition, since the air flow parameter is inherently averaged over a particular area during the measurement process, it is therefore a statistical quantity by its very nature.

Another example is a manufacturing process for drilling a large number of small holes that was optimized by the National Bureau of Standards [6]. In this case, the requirement was to drill 16,844 holes in each of two cylinders that were used to form a punch and die set for forming the perforations in a sheet of stamps. In the intended operation, a pair of matched cylinders was to be rotated together in a synchronized fashion so that pins that were imbedded in the holes on one cylinder were inserted and removed from mating holes on the other cylinder. As long as the holes and pins were properly positioned while the cylinders were rotating, the result was a rotating hole punch. The perforations around the individual stamps were formed as the sheets of paper were passed between the pair of rotation cylinders. As an example of the manufacturing tolerances required, the holes for the dies were 0.043 in. in diameter, 0.320 in. deep, and required a positioning accuracy of approximately 0.0002 in. The holes for the pins used as punches on the punch cylinder had to be located precisely with respect to the matching holes on the die cylinder or the cylinders would be damaged in operation.

While the workpiece positioning requirements associated with this process were reasonably precise, the most difficult task was to avoid breaking drill bits during the hole drilling operations. The previous fabrication procedure had been to change drill bits frequently to avoid the rework problem but removing broken drill bits still required as much as 25% of the fabrication time. The most significant process parameter in this case turned out to be the acoustic emission signature that was produced by a drill bit that was about to fail. Figure 3.12 shows the acoustic emission sensor mounted on a workpiece during a series of drilling tests. An electronic process monitor was fabricated to detect the shift in the acoustic emission signal and used to halt the process before the drill bit could break. In this way the system was able to fabricate the stamp cylinders without requiring rework due to broken bits and tool utilization was improved since it was not

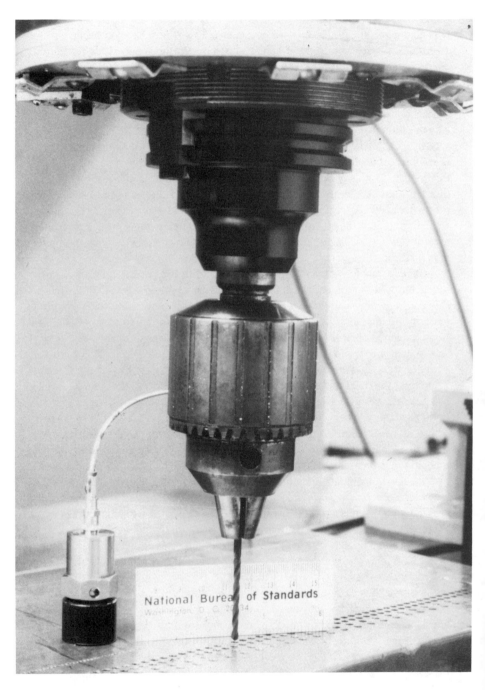

Figure 3.12 Sensor tests for stamp cylinder machining (from Ref. 6).

necessary to perform a tool change as often as had been required with the old fabrication method.

Temperature is a process parameter that is frequently important in manufacturing operations. In some instances, such as heat treating, aging, etc., it is necessary to maintain a specific temperature-time profile in order to obtain the desired result from a process. In other conditions, it is necessary that the temperature be controlled as closely as possible to a specific value because of the system disturbances produced by fluctuations in the thermal environment. One example of the process disturbances that can be caused by temperature variations is the dimensional expansion and contraction of machine components.

Several approaches are available to address the problem of dealing with a process parameter like machine temperature. One technique is to "warm up" a machine prior to performing a critical operation. This can accommodate problems such as spindle growth because the machine is allowed to stabilize prior to finishing a workpiece. However, another alternative is to measure the spindle growth parameter and make the appropriate system adjustments. A second method of dealing with temperature-induced machine motions is to measure the dimensional response of the machine to different thermal conditions and construct an error table that defines the appropriate correction factors. Then, the temperature of critical machine components becomes the process parameter that is used to modify the operation of the system. This technique is used with precision coordinate measuring machines. A third approach is to attempt to maintain a constant temperature environment. This can be accomplished through the use of the room temperature as the controlling process parameter. However, this technique may not be sufficient in those cases in which hot or cold spots exist within a room due to the imperfect mixing of the room air. In this case, the use of a controlled-temperature liquid may be appropriate to control the temperature of the machine. This type of system flows the liquid over the critical portions of the machine in order to establish dimensional stability, and the temperature of the supply liquid becomes the controlling process parameter.

Error Conditions

Error conditions may be defined as those circumstances in which one or more critical process parameters have deviated significantly from their intended setpoint values. (Obviously, error conditions also exist for the noncritical process parameters. However, this situation is not important to the present discussion because the effects of these error conditions are not significant with respect to the process quality.) If no control mechanism exists to correct the parameter excursion or to accommodate the effects of the resultant process shift, then the error sources will produce a set of circumstances that result in unwanted process characteristics.

The effect of the error condition on the process output depends on the sensitivity of the process to the particular combination of events. An error source that produces a significant error condition in one instance may not be influential in other situations. For example, a lathe that is being used to manufacture cylindrical workpieces is relatively insensitive to error conditions that cause deviations in the tool position in a direction parallel to the cylinder axis. This is because the error condition only results in variations in the amount of material that is removed for each revolution of the part and does not affect the final shape of the workpiece. However, on a facing cut, these same error conditions would result in significant flatness errors. In a similar fashion, error conditions that cause the cutting tool to move in a direction perpendicular to the workpiece surface are reflected directly in the contour/diameter quality of the cylindrical part. Also, tool path errors in the semifinish cycle of a milling operation are not serious as long as sufficient material remains to produce the desired finished part (assuming part datums are not disturbed).

One common error source on machine tools is the leadscrew that is used to propel linear slides. Eccentricities in the slide drive system result in a cyclic disturbance being applied to the axis. This results in an error condition that is a combination of linear and angular error motions that have a period that is a function of the pitch of the leadscrew. For a single-axis operation the cyclic errors that occur along the axis path are in the insensitive direction, but the perpendicular errors can degrade the workpiece quality as discussed previously. When two machine axes are involved, the error motion becomes more complex because the phase and amplitude of the various components interact to produce the error motion. This composite error may be parallel to the direction of travel (axial), perpendicular to the direction of travel (lateral), or anywhere in between as shown in Figure 3.13 for a machine with X and Y axes moving at the same speed. This relationship is further complicated when the two axes are moving at different speeds. In this case, the resultant error motion again depends on the amplitude and phase of the various error components. However, rather than having a single frequency sinusoidal error as occurred previously, now the waveform has two frequency components as shown in Figure 3.14.

Another convenient way of representing the error condition that is brought about by the interactions of the error motions of the two axes is through the use of Lissajous patterns. These curves are the paths that are traced by a point (such as the center of a cutting tool), that is simultaneously oscillating in two mutually perpendicular directions. In general the frequency and amplitude are different in the two directions, and the initial phase angle may also be a variable. Figure 3.15 shows how Lissajous patterns are determined from the two orthogonal motions. Figure 3.16 shows a family of Lissajous patterns that can be obtained with equal motion amplitudes but different frequency ratios and phase differences. It should be apparent that if the amplitudes also vary, then curves will be even more complex. The curves in the first row (1:1) of Figure 3.16 correspond to

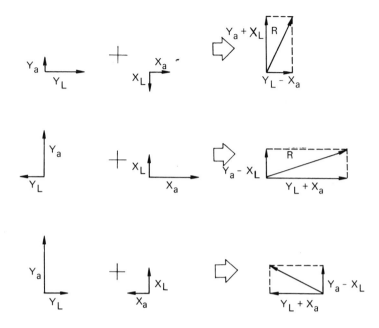

Figure 3.13 Graphical description of error motion interaction on a two-axis machine (axial and lateral errors present on both axes) (courtesy of Martin Marietta Energy Systems, Inc.).

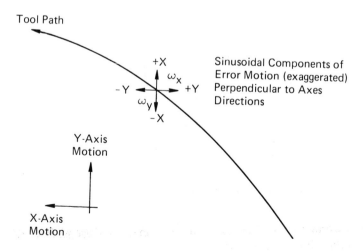

Figure 3.14 Tool path contour that results in unequal axes feedrates and different frequency error motions (courtesy of Martin Marietta Energy Systems, Inc.).

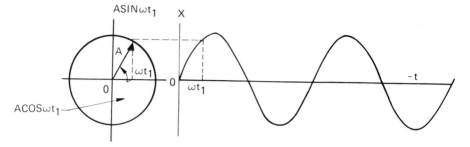

(a) Relationship between sinusoidal waveform and rotating vector.

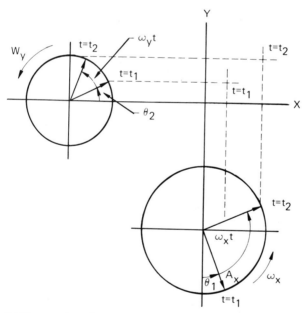

(b) Construction of Lissajous figures from rotating vectors.

Figure 3.15 Determination of Lissajous patterns from two orthogonal motions (courtesy of Martin Marietta Energy Systems, Inc.).

Figure 3.16 Lissajous patterns depicting error motions for various feedrate ratios and initial phase differences but equal disturbance amplitudes (courtesy of Martin Marietta Energy Systems, Inc.).

the error motions that would occur when both axes were moving at the same velocity, as discussed earlier. The other curves represent the error motions that would occur for a variety of other operating conditions when following a nonlinear tool path.

As shown by the curves in Figure 3.16, it is often difficult to measure directly the process disturbance and implement a real-time correction technique. Instead, if a compensation method is employed, it is more appropriate to isolate the error source into its basic components (the individual axes error motions) and perform corrections for these simpler components. An alternate approach is to reduce the error source to an acceptable level.

Because of the complex manner in which error sources can interact and the varying degrees of sensitivity that a process may exhibit to these error conditions, it is inappropriate to halt a process solely due to the existence of one or more error conditions. Instead, it is necessary to evaluate, ahead of time, the potential error conditions with respect to the effect that these errors would have on workpieces at different phases of the fabrication cycle. In addition, the potential implications for the finished part must be assessed. Then, an operating strategy can be formulated in accordance with the response that is most appropriate at this particular point in the manufacturing process. Depending on the circumstances, the most suitable reaction to the error condition may vary from ignoring the situation to initiating corrective actions or halting the process. A practical error recovery capability will enable a system to restart the manufacturing cycle without having to "purge" the system and start over from the beginning of the process. The automation of this procedure for error detection and recovery is a step in the direction toward an "intelligent system" that is capable of responding automatically to a variety of perturbations in the manufacturing environment.

Figure 3.17 shows a computer-controlled coordinate measuring machine design that has been proposed by Lawrence Livermore National Laboratory [7]. The design of this machine incorporates a temperature-controlled oil shower and a variety of other error correction techniques that are intended to avoid the problems that are typically encountered when attempting to perform very high accuracy dimensional measurements. This type of "Y-Z" measuring machine (a machine which has Y and Z linear axes) typically is used to inspect parts that are fabricated as figures of rotation. It is also capable of inspecting some features on other types of parts, but it is not necessarily as easy to use in these other applications.

The typical measurement procedure used with this type of inspection gage is to align the axis of a part to the center of rotation of the rotary table. Then, workpiece contour measurements can be performed at various locations on the part surface through the use of the rotary axis. This is accomplished by positioning the workpiece and the Z-axis gaging probe at the proper locations to record the deviations in the part diameter as the part is rotated. Spherical contours are measured by aligning the centerline of the spherical surface to the rotary table

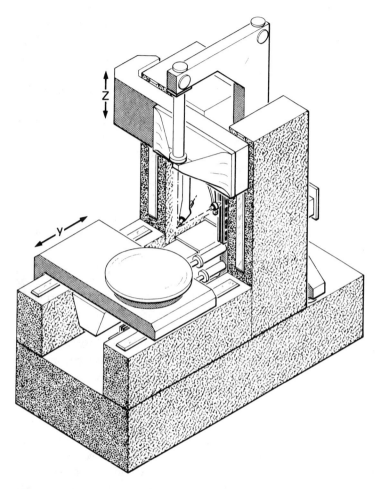

Figure 3.17 Y-Z inspection machine (from Ref. 7).

center of rotation and recording the deflection of the gaging transducer as the part is rotated and the Y-axis slide is held in a fixed position. In addition, non-spherical surfaces can also be inspected if the appropriate linear axis motions are coordinated with the rotation of the workpiece. Also, continuous longitudinal or spiral sweeps may be made on contoured surfaces, although it is necessary again to move the appropriate axes under precise positioning control to assure the coordination of the machine's axes.

The displacement transducer used to gage the workpiece is a ball-tipped, linear variable differential transducer (LVDT) that is attached to the machine's Z axis.

In addition, the axis of the LVDT is frequently located at a 45 degree angle with respect to the Y and Z axes. This means that the part can be inspected anywhere between the horizontal and vertical planes as long as the proper correction is made for the cosine error (the error that is produced by the displacement transducer being at an angle to the part surface).

The machine is designed to use laser interferometers and helium-shielded pathways for displacement measurement accuracy. The Z-axis interferometer is located so that the extension of its measurement axis passes through the center of the tip of the measurement stylus, thereby avoiding errors due to the possible angular pitch motion of the Z axis. The Z-axis roll and yaw motions are of secondary importance since this error motion is in an insensitive direction, as discussed earlier. Two laser paths are used on the Y axis to detect the axis displacement and pitch angular motion. The difference in the readings between the two axes displacement measurements is available as an input to a piezoelectric-crystal-controlled servo system that corrects the slide motion error by raising or lowering one end of the Y-axis slide.

Slide straightness errors are measured by using LVDTs that are in contact with straight edges that are mounted parallel to each slide. The straightness errors on each slide cause deflections in the approprite LVDT and the control system uses this information to correct the displacement of the opposing axis. This compensation technique is designed to counteract the effect of the error motion.

The machine design also incorporates a metrology reference frame (see Figures 3.18 and 3.19) that is independent of the machine base. Since the machine base cannot be made infinitely stiff, it is possible for deflections to occur due to variations in slide position and other system disturbances. In order to realize the maximum measurement-accuracy potential of this machine it is necessary to decouple the metrology system from the deflections of the machine base. This insures that the forces on the metrology system are constant during normal operating conditions. To accomplish this objective, the metrology frame is supported on kinematic mounts that are located inside the machine base. In addition, the plane of the kinematic supports is coincident with the neutral bending axis of the machine base. This minimizes the effect that the deflection of the machine base has on the accuracy of the workpiece measurements.

Variations in the thermal environment of the measuring machine and the workpiece are very important in this application. Therefore, this machine is designed to be showered with 40 gallons per minute of oil that is controlled to a constant temperature. The potential sources for temperature-related errors that are associated with this system include the laser, the slide drives, the operator, and the control system as well as any peripheral equipment that may generate heat.

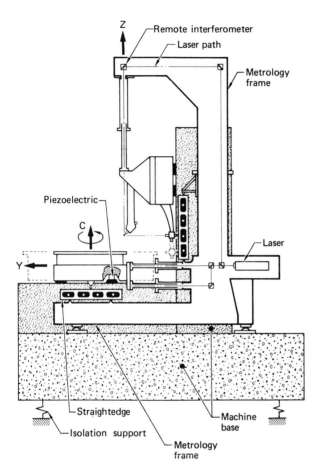

Figure 3.18 Y-Z machine cross section (from Ref. 7).

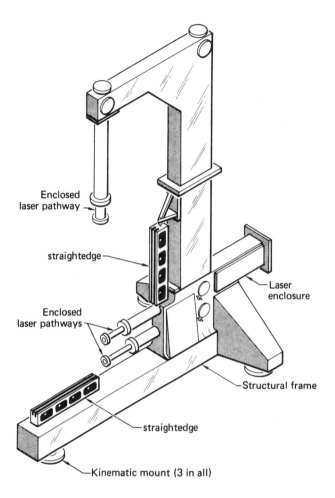

Figure 3.19   Metrology frame for Y-Z machine (from Ref. 7).

Operating Strategy

In the example just presented dealing with the design of a Y-Z measuring machine, the operating strategy is to depend heavily on the mechanical accuracy of the machine. The design goal is to make the basic machine as accurate as possible and to correct, in realtime, the motion of the machine's linear axes for the positioning errors that arise from slide straightness and angular motion errors. While this may be sufficient for a measuring machine it only begins to scratch the surface of the problem for a metal cutting machine. As mentioned previously, there are a number of other error sources that occur in the machining environment that are not present in the inspection world. While it is desirable to have an accurate machine, it is not necessarily sufficient, due to the likelihood of errors being introduced by variations in material, fixturing, tooling, etc. In this case, focusing exclusively on machine accuracy results in "wasted" resources since the major portion of the problem may be ignored in pursuit of a minor portion of the system errors.

If it is assumed that the errors in a manufacturing process are independent of each other and have root mean square (rms) amplitudes equal to RMS(i), then the combined effect of these errors can be expressed as

Total RMS error $=\{\Sigma \ [RMS(i)]^2\}^{1/2}$

In this case, error sources that can be made relatively small with modest effort should be reduced, rather than allow all errors to be comparable to the largest error [8]. If N error sources are allowed to have the same amplitude, E, then the above equation gives a total error of $N^{1/2} \times E$. Alternatively, if all but the one error is reduced to 10% of the largest error, then the total error is

Total RMS error $= \{[1 + 0.01(N - 1)]^{1/2}$

For N equal to 10 the two values given by the preceding equations are 3.16E and 1.04E respectively, which is a significant difference. However, it may also be shown that once an error is reduced to 10% of the dominant errors, that there is little benefit from reducing it further.

Figure 3.20 shows a two-axis, computer-controlled lathe that is used as a pilot facility for developing deterministic manufacturing techniques for potential use in the production shops at the Oak Ridge Y-12 Plant [8]. This machine is equipped with ball–nut leadscrews, electric drives, roller slide bearings, an air bearing spindle, laser interferometer position feedback for the linear axes, an on-machine gaging system and an automatic tool changer. The machine is capable of fabricating an 8 in. diameter hemispherical contour with a form accuracy of better than 0.0001 in. when using a diamond tool to optimize the metal-cutting performance and a solid-aluminum mandrel for a workpiece (to avoid part deflection or deformation problems) [9]. In addition, a significant portion of this error can be attributed to the size and shape discrepancies in the diamond tool. (An alternative technique for establishing tool path accuracy is to replace the cutting tool with an LVDT and record the gage readings as the machine is used to trace the surface of a part with a known form such as a disk or tooling ball.) While this machine can be shown to have an excellent tool path accuracy when performing in an ideal environment, in a production operation it may not be capable of achieving this same level of contour accuracy. This apparent performance degradation is not necesarily the fault of the machine as much as it may be due to other factors such

Figure 3.20 Two-axis CNC lathe (from Ref. 9).

as the suitability of the tooling and machinability of the workpiece. While it is possible to attempt to use the machine in a manner that overloads its capacity and thereby cause performance problems, there are other complicating factors that are beyond the control of the machine tool. The required workpieces may not be fabricated from stress-free materials, it may be necessary to secure the parts in a fixture and machine them on both sides, tool wear can be a negative influence, and parts are usually required to be made to a specific size (which requires an accurate technique for establishing the reference datum for the cutting tool), and it is often necessary to locate one or more surfaces in a specific orientation with respect to a part datum. All of these factors influence the quality of the final product but are relatively unrelated to the tool path accuracy of the machine tool.

In addition to having the capability for generating a high-accuracy tool path, this machine also utilizes in-process monitoring techniques. This is done to estimate the quality of the workpieces prior to their removal from the machine so that process excursions can be detected as rapidly as possible and perhaps permit a problem feature on an affected workpiece to be corrected. If no anomalies are detected, then this information can be used to justify a sampling process for part certification as well as to provide historical information on the conditions that were encountered during the manufacturing process and the resultant part quality.

Figure 3.21 shows a touch-trigger probe that is used to provide dimensional information concerning the machine and its workpieces. This device is a three-dimensional probe that has internal electrical contacts that open when the stylus is deflected. The probe is used by moving the it toward a surface of interest and recording the position of the appropriate axis at which the electrical contact is opened. Two gaging runs are used for each probing location. The first approach is made at a relatively high speed to obtain a rough approximation of the surface location, while the second run is made at a slower speed to gain more accurate information. In general, length measurements are conducted using the same contact location on the stylus for all measurements so that the difference in the machine axes locations at which the electrical contact signal occurs is equal to the feature length (assuming no errors in the measurement of the axis travel). However, if diametric or other measurements involving multiple contact points on the probe stylus are employed, then a mastering technique must be used to calibrate the gaging system.

The data defining the workpiece and machine dimensional characteristics, as well as the self-calibration cycle used by the gaging probe, are accumulated automatically under the control of the machining part program. This manufacturing process knowledge is sent to a data base on a host computer where the information is sorted according to the particular operation. The data are also used to construct control charts that define valid operating ranges for the process. (More information on control charts is contained in a subsequent chapter on statistics.) The upper and lower control limits for each of the monitored parameters are down

Figure 3.21 Touch-trigger probe used for on-machine gaging (courtesy of Martin Marietta Energy Systems, Inc.)

loaded to the shop floor computer. This enables the monitoring system to use up to date information and provides accurate, rapid detection of a manufacturing problem.

Other process characteristics that are monitored on this machine include tool wear, vibration levels, temperature, fit between the workpiece and the fixture, servo system characteristics and the sequence of commands issued to the control system (control system status). The operational strategy for this machine is to use the dimensional information obtained through the on-machine gages to characterize the part quality at the semifinish (just prior to the last machining pass) and finish machining stages. Offsets are made to the machining process prior to the last pass if the in-process measurements predict that a significant error will result unless corrective action is taken. Examples of error conditions that would result in the process being halted are a significant excursion in coolant temperature, excessive servo system following error or imbalance, excessive tool wear or tool breakage, or a significant shift in the gage calibration factors. Other types of error conditions generate a flag to the monitoring system but do not require the process to be shut down. Examples of these types of conditions include excessive values for most of the monitored parameters during a rough machining operation, although tool breakage and servo system problems are serious errors at any point in the machining cycle.

The Y-Z measuring machine and the two-axes lathe discussed above are designed to employ two different approaches in order to obtain the desired performance characteristics. The design of the gaging machine was based on the utilization of sensors to detect position errors due to angular, straightness or axial error motions. Then, the control system is depended on to initiate the appropriate corrective action to implement the necessary physical adjustments via additional inputs to a closed-loop servo system. The second machine is not expected to maintain the same degree of tool path accuracy as the first because there are other more significant error sources that must be accommodated. In this case, assuming no catastrophic occurrences such as tool breakage, the corrective actions are implemented only for the final machining pass and are based on the performance that was observed after the semifinish machining pass. (Some features, such as axial-slide position, spindle speed and coolant temperature, are continuously controlled as a part of the basic machine control system.) Also, additional actions may be taken at the appropriate time if a significant excursion occurs in one or more of the monitored process parameters.

A third method that can be used is to map the machine errors and use a computer data base to provide the information needed to correct the machine errors. This technique is an extension of the leadscrew correction option that is available with most computer control systems. It is a viable approach as long as the error conditions that are encountered during a given operation are covered by the information in the data base and the error detection/correction can be performed in sufficient time to ensure an adequate response to the changing environmental conditions. The National Bureau of Standards (NBS) has demonstrated this error mapping approach by creating a machine independent coordinate system that is based at a single point [10]. Measurements made relative to this reference frame are transformed into the coordinate system of the measured objects using rigid body kinematics. Error terms associated with the mechanical condition of the machine are measured on a cubic lattice and stored as matrices that are used to correct the data during the measurement of a workpiece.

The NBS first used the "classical" manual measuring machine, shown in Figure 3.22, which had a measurement volume of 48 × 24 × 10 in. to demonstrate this error compensation technique. This machine was retrofitted with laser interferometers to provide position feedback of the axes displacements and a minicomputer was added to enable automatic operation. Prior to the retrofitting it had a worst case accuracy of several thousands of an inch along a diagonal of the

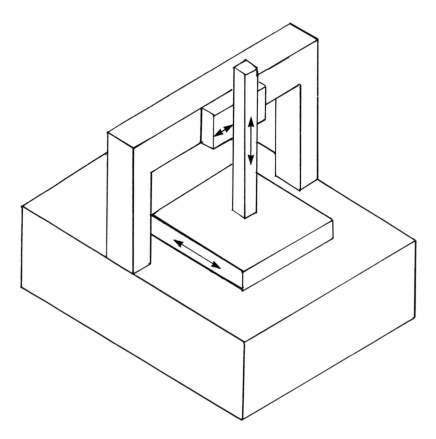

Figure 3.22   Schematic diagram of the manual measuring machine.

measurement volume. Figure 3.23 shows a plot of the roll error of the machine table as a function of its X-Y position. Use of the computer data base and the appropriate techniques to eliminate nonrepeatable error sources enables this machine to have a measurement capability of about 50 μin. (1.3 μm).

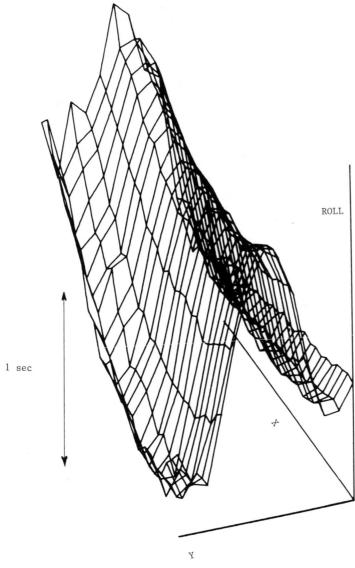

ROLL

1 sec

Y

Figure 3.23 Roll motion of the measuring machine table (courtesy of the NBS).

Subsequent activities at the NBS have evolved into the development of a general methodology for machine tool accuracy enhancement through the use of error compensation [11]. This technique involves a general mathematical model, which relates the error in the relative locations of the cutting tool and workpiece to the errors of the individual elements of the machine tool structure. Following calibration checks of the machine tool, the errors of the individual machine elements are decomposed into geometric and thermally induced components. Based on this calibration data, the empirical models are obtained using a least-squares curve-fitting technique. Then, a flexible, modular and structured computer software model can be used to compensate for the machine errors in real time. This compensation system monitors the temperatures on the machine tool structure and the nominal axes positions. Then, this information is used to define the compensation actions that are needed to correct the predicted machine errors. Additional details on this monitoring technique are included in the process monitoring chapter.

Summary

The deterministic manufacturing approach does not really represent a new technology for the manufacturing community. Instead, the methods described are a reemphasis of sound engineering practices that have been utilized to a certain extent from the beginning of manufacturing activities. These formalized techniques may be more widely employed now owing to the availability of new sensor systems and the increased use of computer systems, but the basic idea is no more complex than the use of a speedometer in an automobile. A speedometer is used to provide an indication of the speed at which an automobile is traveling so that the control system (the driver) is able to maintain the quality of the process (arrive at the destination safely without receiving a traffic citation for speeding). Because a direct measurement of the car's velocity is difficult to achieve economically, the rate of rotation of the drive train is used as an approximate process parameter (the true velocity is also dependent upon the effective diameter of the tires). The main difference between the deterministic manufacturing concept and closed-loop feedback control systems can be demonstrated through the extrapolation of the previous example. This difference occurs in the use of the process information that is obtained from the various process sensors. For example, an automatic, speed-control system for an automobile employs an empirical approach for maintaining vehicle speed. It measures the drive train angular velocity and adjusts the engine throttle position to the required location to keep the speed at the present value, in spite of variations in the terrain. The degree of precision achieved is dependent on all of the system variables but the results are usually accepted as being correct. However, the process quality usually is not recorded, so that there is no "internal proof" that the speed limit has not been exceeded at some time or that the maximum recommended value for engine rpm has not been exceeded. Of course, external spot checks of system performance can be performed through the use of radar or other means. In contrast, the deterministic manufacturing approach does not just try to control system variables as a means of producing the desired result. Instead, this technique attempts to verify, with a high level of confidence, that success has been achieved. In the automobile speed-control system example, the vehicle speed and engine rpm could be monitored with a computer to provide independent verification that the speed limit was not exceeded and the engine was not operated at an excessive speed.

The deterministic manufacturing philosophy is not new; at the same time, the successful implementation of the technique is not a trivial task. The identification and analysis of the key process parameters can be quite a formidable task in some instances. In addition, obtaining an adequate system model for use with the critical process parameters can be a complex endeavor and the only practical approach in some situations may be to utilize a statistical model.

The goal of the deterministic manufacturing method of product fabrication is to demonstrate statistically that a process is capable of meeting the desired goals with a high degree of confidence. Then, the key process parameters of the manufacturing operation are monitored to certify that the process is functioning as expected. This is done by showing that a system is operating within statistical limits that have been historically proven to produce the desired product to a quality level that is a small fraction of the allowed tolerance. The result is a high degree of confidence that the specified tolerances are being achieved. In addition, in the event that a system malfunction does occur, it will be detected immediately and not jeopardize the validity of the process certification operations. This enables a sampling procedure to be employed to verify that the process is still functioning as expected. However, the use of continuous in-process monitoring precludes the necessity of periods of 100% postprocess inspection that are normally required following the detection of faulty product in a conventional sampling system. This advantage is achieved because the problem is detected before the product gets to the final inspection stage of operations. Therefore, except for prove-in periods, no faulty products should reach the sample inspection stage, and the products that are inspected are only used to establish the confidence limits for the process.

References

1. Torque monitoring system does more than tighten bolts, *Production Engineering 34*: 2 (1987).
2. Temperature and Humidity Environment for Dimensional Measurement, American Society of Mechanical Engineers, New York, ANSI B89.6.2, 1973.
3. N. D. Woodall, Part Stability and Fixturing Techniques for Precision Machining, Union Carbide Corporation, Oak Ridge Y-12 Plant, Oak Ridge, Tennessee (Y/DA-7177 (1977).

4. O. T. Gibson, Workpiece Fixturing Using Rubber Compounds, Union Carbide Corporation, Oak Ridge Y-12 Plant, Oak Ridge, Tennessee Y/DA-6740 (1976).

5. W. H. Rasnick and M. L. Shell, Impregnation of Porous Graphite for Flow Modification, Union Carbide Corporation, Oak Ridge Y-12 Plant, Oak Ridge, Tennessee, Y-SC-7 (1971).

6. R. P. Bergstrom, Drilling precision stamp cylinders at NBS, *Manufacturing Engineering, 96*: 10 (1984).

7. J. B. Bryan and D. L. Carter, Design of a New Error-Corrected Coordinate-Measuring Machine, Lawrence Livermore National Laboratory, Livermore, California, UCRL-82450 (1979).

8. The Oak Ridge Y-12 Plant is Operated by Martin Marietta Energy Systems, Inc. for the U.S. Department of Energy Under Contract DE-AC05-84OR21400.

9. W. E. Barkman and L. M. Woodard, Upgrading a Production Machine Tool for Precision Machining, Union Carbide Corporation, Oak Ridge Y-12 Plant, Oak Ridge, Tennessee, Y-2264 (1981).

10. R. J. Hocken, Three Dimensional Metrology, National Bureau of Standards, Washington, D. C.

11. M. A. Donmez et al., A general methodology for machine tool accuracy enhancement by error compensation, *Precision Engineering 8*:4, (1986).

CHAPTER 4

Statistical Control of Manufacturing Processes

Introduction

The introductory statistics chapter covered preliminary material dealing with basic statistical concepts. This discussion provided a demonstration of the relatively uncomplicated nature of the mathematical calculations and concepts that can be employed to provide an objective description of a manufacturing operation. Characteristics such as the mean, variance, standard deviation, and probability density functions were presented in light of the manner in which these statistical entities could be used to describe populations of measurements. The subsequent chapter described deterministic manufacturing methods in a somewhat general fashion. This presentation described a manufacturing philosophy that depends heavily on what is essentially common sense, which is bolstered by the use of statistics, to control the quality of a manufacturing operation. However, no explicit information was presented on how particular statistical techniques were used to achieve a higher level of quality control. The present chapter shows some statistical methods that can be employed beneficially to obtain an accurate model that describes the condition of a process. Once this analysis has been accomplished, the quality of the operation can be estimated accurately and therefore controlled. (Assuming it is permissible to perform process parameter adjustments as a means of achieving improved system performance.)

One set of circumstances that is often a concern with certification programs is the question about the occurrence of multiple defects in the output of a manufacturing process. This situation involves a subject called *conditional probability*. It deals with instances in which one event is known to have occurred and the problem is to determine the likelihood that another event also has happened. If the events of concern are independent events, then the probability of the second event occurring, given the fact that the first event has already occurred, is the same as the probability for the second event occurring by itself. The probability of both events occurring is the product of the probabilities of the individual events. On the other hand, if the events in question are not independent of each other, then the occurrence of one event influences the likelihood of the occurrence of the other event. Interest in these topics leads to an area of statistics called *acceptance sampling*, where it is assumed that a certain percentage of defects exist and information that defines the probability of locating these defects via a sampling plan is required.

One technique for estimating the quality of a manufacturing process involves the use of sampled data for the determination of confidence limits. These statistical entities are set up through a mathematically sound procedure that provides assurance that everything is acceptable as long as certain features remain within prescribed limits that are specific to a particular process. Statistical control charts are an extension of this approach. In this instance, the statistical information is plotted in a specified sequence that allows patterns in an operation to be detected (via process parameter measurements) before a catastrophic failure occurs. Error budget analysis is another statistical tool that permits modeling of a process for design and in-process control. The following sections of this chapter will discuss these areas in more detail and give specific examples of practical applications. In some circumstances, it may be that the depth of the discussion of statistical calculations or the use of equations involves excessive detail for certain readers. If this is the case, it is suggested that the beginning of each section of this chapter be read but that excessive time not be spent worrying about an in-depth understanding of the statistical details. (You can always hire a statistician to worry about the pertinent specifics for a given application.)

Random Sampling for Acceptance

Chapter 2 discussed the use of various statistics, as well as probability distributions, as a technique for evaluating the characteristics of a process. While in many cases it is possible to accumulate an extensive reference set of data for use in the analysis of a manufacturing operation, this is not always practical from a time or cost standpoint. One example is the situation in which a customer is purchasing a group of 100,000 parts from a vendor and wishes to estimate the quality of the lot without certifying each individual piece part. A similar situation exists when the testing of a component destroys the unit's usefulness. For example, testing the tensile strength of a bolt is accomplished by stressing it until it fails. If the assumption is made that a sample of observations of the production process is a random sample of an appropriate population, then a viable reference distribution can be obtained from a smaller group of data. Stated more simply, this means that the characteristics of the sample are similar to the characteristics of the population. Therefore, the quality of the sample can be used to estimate the quality of the process.

The general procedure in acceptance sampling operations is to select a random sample of n articles from a larger population N. The lot is then accepted on the basis that not more than an allowable number, c, of items in the sample is found to be defective. Otherwise the lot is rejected. Thus, a sampling plan for a given population is dependent upon both the sample size and the number of defectives permitted. In some instances, multiple sampling plans may also be utilized.

If items cannot be tested without destroying the usefulness of the product, then rejection of a lot means that the population is effectively scrapped. Alternatively, if nondestructive testing is available, then rejection of a lot based on the sampling plan offers the option of performing 100% inspection on the remaining members of the group. Another option is to offer the material at a substantial discount and shift the burden of 100% inspection to the user. A lot in which the defective members have been replaced with acceptable products is called a rectified lot.

The probability of being able to accept a lot, $p(A)$, based on a sampling procedure, may be obtained by summing the probabilities of finding each of the possible number $(0, 1, 2, \ldots, c)$ of defective units in the sample. (The probability of rejecting the lot is the summation of the probabilities of finding each of the $c + 1$ to n numbers of defective units, or $1 - p(A)$.) Letting $p(x)$ be the probability of having x unacceptable items in a sample gives

$$p(A) = \Sigma\, p(x)$$

Obviously, the value for $p(A)$ depends on the population size N, the sample size n, the allowable number of out of tolerance parameters, and the number of defects, θ, that exist in the population. Rewriting the above equation to show the dependence on the percentage of defects in the population gives

$$p(A|\theta) = \Sigma\, p(x|\theta).$$

It should be apparent that $p(A \mid 0) = 1$, since if there are no defects the lot is sure to be accepted (assuming the nonexistence of the real-world problem of testing errors). Also, as the number of defects θ increases, the probability of acceptance decreases until $p(A|1) = 0$.

In the acceptance sampling by attributes method discussed above, the only information about a product that is used is whether or not the item meets a specification. This technique is applicable when go-no go type gages are used for certification. However, if the actual value of a product characteristic is measured in the inspection process, then only a fraction of the available information is being utilized by this procedure. A better approach is to use the actual measurement values to obtain a more sensitive sampling plan, which generally requires smaller sample sizes than the attribute sampling method.

As an example of a one-sided acceptance plan that is based on the mean of the samples, consider the tensile testing of bolts that was mentioned earlier. Assume that a bolt is considered to be acceptable if it has a yield point of 50,000 psi or higher. In addition, it is known from past experience that the standard deviation of the yield points is about 2,500 and that this characteristic approximately exhibits a normal distribution. Then, an acceptance plan can be set up where the lot of bolts is accepted if the mean of a sample of size n exceeds some number k (where n and k are to be determined). Otherwise the lot of bolts is rejected. (Sampling plans may also be based on the variance of the sample lot. The objective in each case is to assign a level of confidence to the assessment that the product meets the required specifications.)

Since this acceptance criteria is based on a statistical sampling approach, there is a certain risk to the producer and consumer that a particular lot will be inappropriately accepted or rejected. The risk to the consumer is that a bad lot will be accepted, while the risk to the producer is that a good lot will be rejected. If the producer and consumer agree on their respective risk factors for specific values of mean yield strength from a sample lot, then a sampling criteria can be established that defines the values of the parameters, n and k.

Assume that a sampling plan has been established in which it is decided that if the mean yield strength of the sample lot is 55,000 psi (approximately two standard deviations above the minimum acceptable value), then the probability of accepting the lot is 0.01. (The risk to the consumer is 10% that this lot will be accepted.) In addition, if the lot mean is 57,000 psi, the probability of accepting the lot is 0.95. (The risk to the producer of this lot being rejected is 5%.) From the discussion of the normal distribution in the earlier statistics chapter, it is assumed that the sample mean has approximately a normal distribution as defined by

$$N\left(55,000 \; \frac{2500}{\sqrt{n}}\right)^2$$

where the lot mean is 55,000 psi and the standard deviation is 2,500 psi. Using the consumer's and producer's risks, and a table of the percentage of significance points for a normal distribution, it can be calculated that a sample size of 14 and a sample mean acceptance value of 55,876 psi are required. [1] (The details of this calculation are not appropriate for inclusion in this text.)

This type of sampling procedure could also be used with a system that employs deterministic manufacturing techniques for maintaining product quality. However, instead of accepting products based on the results of the sample lot, the sampling procedure is used to verify the integrity of the in-process monitoring procedures used to certify the process output.

The discussion of acceptance criterial and risk factors implies an observation about inspection/sensing operations which can sometimes be forgotten. An unfortunate fact of life is that measurements are not perfect. There is some inspection error associated with each inspection process. (While we can usually count how many units are in a given lot, it may be difficult to agree upon the exact size of each unit.) The amount of the total manufacturing tolerance that is consumed at the measurement stage depends upon the repeatability and accuracy of the inspection operation. This system tax cannot be avoided and must be considered during process analysis/enhancement activities.

Confidence Limits

The previous section covered the estimation of the parameters of a population of manufactured products through the statistical analysis of the characteristics of a small random sample. While the results of these techniques are called estimates, it is assumed that the information is a reliable indicator of the status of the entire population. This assumption is based on a measure of confidence that is determined by the procedures used to arrive at the statistical values. However, it is not unusual for the average of some measurable characteristic associated with a given sample to differ from the true average of the same characteristic for an entire population due to the natural uncertainties that arise from fluctuations in the sampling process.

For sample sizes equal to or greater than 30 units, the sampling distribution of many statistics will approximate a normal distribution. In this case, it can be expected that a sample statistic will be found in a range equal to the mean of the sampling distribution plus or minus two times the standard deviation about 95.45% of the time. Table 4.1 shows these confidence levels as a function of the multiplication factor that is applied to the standard deviation to determine the allowable range. Similarly, the sample mean plus or minus 1.96 times the standard deviation will provide a range that includes the population mean 95% of the time. [2].

As might be expected, the smaller the sample size, the larger the variation between the sample average characteristic and the population average characteristic. For sample sizes less than 30 units, the approximation used above becomes less accurate and the estimation procedures must be modified. In this event, the t distribution mentioned previously is employed. If the sampling is done randomly from a statistically controlled population (with respect to the measured variable) that has an approximately normal distribution, then a range can be calculated that is determined by the specific confidence limit that is desired. The following equation performs this task through the

$$\overline{X} \pm Ls$$

Table 4.1 Confidence Levels for Various
Standard Deviation Multipliers

Confidence Level	Multiplier
99.73%	3.00
99%	2.58
98%	2.33
96%	2.05
95.45%	2.00
95%	1.96
90%	1.645
67.28%	1.00
50%	0.675

Source: Ref. 2.

use of the average and standard deviation of the n measurements. L is determined from Table 4.2 for different sizes of samples and a particular confidence limit [3].

For example, the average weight of the marbles in a 100,000 unit population might be estimated by weighing a sample of 10 marbles and expressing the result as 10 grams plus or minus 0.1 gram. But this does not provide any confidence level information. As an example of a more informative approach, consider a group of randomly selected measurements of the diameter of replacement pencil lead for mechanical pencils. Assume six samples are taken randomly from a homogenous population with a nearly normal distribution. Then, through the use of Table 4.2 it can be expected that 95% of the time the average pencil lead diameter will fall within a range specified by the average of the six sample measurements plus or minus 1.050 times the standard deviation of these same measurements. (The statement that "the probability is 95% that a factor falls within specific limits" is meaningless because the factor either does or does not occur within a specific range.)

Control Charts

As discussed in the previous sections, measurements are frequently made of certain workpiece features or process parameters in repetitive manufacturing operations. Then, these characteristics are used to describe the quality of the process from a statistical viewpoint. For instance, in most cases it is desirable that the mean value of the process characteristics be centered within the tolerance specification

Table 4.2 Factors for Calculating Confidence Limits for Averages

Sample Size	Confidence Limits		
	90%	95%	99%
4	1.177	1.591	2.921
6	0.823	1.050	1.646
8	0.670	0.836	1.237

Source: Ref. 3.

for the product and that the standard deviation be small enough to satisfy the through-put requirements. While it may be desirable in some situations to bias a process so that the mean is not centered within the product specification, this is usually done only because of complicating factors that cannot be readily resolved in another manner.

In a normal manufacturing environment, it is not an unusual situation for the values of the process parameters to change somewhat from one measurement to the next (due to the lack of perfection in the manufacturing and gaging operations). Although a goal of a particular organization may be to refine a manufacturing operation to the point that the process parameter variations are as small as practical, from a statistical standpoint, these variations are considered to be a natural characteristic of the process behavior. If a stable operation is assumed, the result is that the successive process-parameter measurements exhibit the same characteristics as if they were obtained at random from a population of measurement values having a particular mean and standard deviation. Therefore, if $x(1)$, $x(2)$,. . ., $x(k)$ measurements are taken on a sequence of k successively selected units, then these samples of measurements should behave like independent random variables which have identical distributions. In addition, the $x(i)$ values should behave like a random sample of size n from a larger population with some distribution. Avoiding statistics terminology, this means that accurately characterizing the sample provides a realistic appraisal of the process output.

A process is under statistical control if the measurements made in a sequential manner act like a series of independent and identically distributed random variables. This statistically in-control process is one in which the process variations are no greater than those that can be attributed to chance. Unfortunately, just because a process is operating in a condition of statistical control does not mean that it is also capable of producing a product to the required level of quality. However, it should be noted that describing an operation as lacking statistical control (an out-of-control process) has nothing to do with the conventional controls terminology used to describe the operation of servo systems. An out-of-control servo system offers the possibility of catastrophic results for the manufacturing equipment as well as being a safety problem. In contrast, a process which is not in a state of statistical control may be operating normally but it is being disturbed by a characteristic of the incoming feed material that is unrelated to the stability of the manufacturing equipment.

The acceptance sampling procedure discussed previously provides a technique for determining when a process is producing "good" or "bad" products. However, it does little to show the presence of trends or to indicate the presence or lack of statistical control. The control chart overcomes this shortcoming by providing a useful technique for presenting data in groupings with respect to time, place, source or other considerations so that a lack of statistical process control can be readily detected. In addition, the control chart may also be used to judge the effects of various attempts to improve the quality of the manufactured product. (A manufacturing operation that is in a state of statistical control may still have ample opportunity for improvements in quality.)

Since it is rarely economical to inspect all characteristics on every unit in a batch of manufactured items, a manufacturer frequently takes a small sample of objects from the process stream and uses this smaller lot to characterize the condition of the process. Control charts are one method used for detecting and analyzing fluctuations of measurements made on successive samples taken randomly from a particular process. If this technique is employed during process start up, when the goal is to reduce parameter variations and attain a condition of statistical control, then the control chart is called an *empirical control chart*. An additional use for a control chart is to monitor the condition of an existing process which is assumed to be operating under statistical control. In this instance, it is assumed that the process mean and variance are established and that the main interest is in detecting excursions in the process rather than optimizing the operation. This type of control chart is termed a *theoretical control chart*.

The theoretical control chart is used for a manufacturing process that is operating under statistical control with respect to specific statistical entities, such as the average value of a monitored parameter. The goal in this situation is to detect variations in the process that are caused by a shift in process parameters. This includes the effects of those parameters that are not otherwise observed but which still influence process quality. The control chart is constructed so that if the process control is disturbed by a significant amount, then the monitoring technique will have a high probability of detecting the excursion. As a hypothetical example, consider a manufacturing process for the production of billiard balls that are to be used in professional billiards tournaments. Since the uniformity of the balls has an effect on the outcome of the game, the diameter of the balls is a critical dimension. It is assumed that the manufacturing process is operating in statistical process control with respect to a ball diameter measurement. This gaging operation is performed at the end of the production line from a random sample of size n. A group of statistic values, such as average diameter or range of diameters, calculated for each of the successive samples will behave like independent random variables having identical distributions. Because the process is operating in a condition of statistical control, this common distribution will have a mean and standard deviation that are known from historical data. For most manufacturing operations, a control statistic (such as the average ball diameter) has a nearly normal distribution, and the probability that this particular sample statistic will fall within specific limits can be determined. If a three sigma control chart is used, as shown in Figure 4.1, then the probability is 99.73% that the statistic of interest will fall within the range specified by the mean plus or minus three standard deviations. Similarly, the probability of this statistic falling outside of this range is less than 0.3%. The probabilities associated with ranges determined with other multiples of the standard deviation can be determined from the confidence level chart shown previously in Table 4.1.

Figure 4.1 contains a center control line (CCL) that is defined by the process mean and represents the historical reference set for this process characteristic. The upper and lower control lines (UCL and LCL) also represent the historical performance of a process because they are determined by the mean plus or minus three standard deviations, respectively. Since the control chart data is plotted in time order, it displays the normal variations in the process in a manner that is recognized easily. In effect, a graphical comparison is made of the process performance against a reference distribution that is based on historical information. As long as the process in under statistical process control (which means that it is operating the way it is supposed to, given its environment), and the control parameter is normally distributed, then 99.73% of the samples will fall within the control limits. Alternatively, this also gives a mechanism for determining when a process is out of control. If this occurs, it is necessary to locate the cause of the disturbance and rectify the situation so that the process can be returned to a condition of statistical control.

Two important statistics for use with control charts are the sample mean, which describes the average condition of some characteristic, and the sample sum which provides insight into the variation within a sample. For a sample of size n, if the population mean, μ, and standard deviation, σ, are known for items being manufactured under a condition of statistical process control, then the control limits for establishing a three sigma theoretical control chart can be calculated as shown below. When the sample mean is used as the statistic that is recorded on the control chart the limits are

$$UCL = \mu + \frac{3}{\sqrt{n}}\,\sigma$$

$$CCL = \mu$$

$$LCL = \mu - \frac{3}{\sqrt{n}}\,\sigma$$

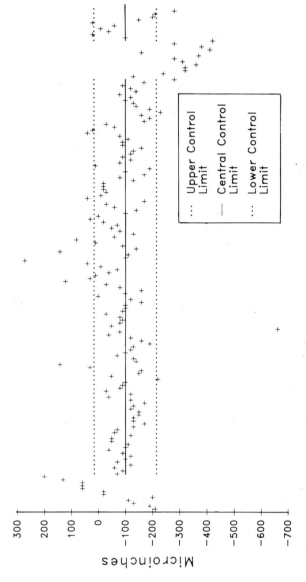

Figure 4.1 Control chart for a three sigma process (courtesy of E. F. Babelay).

If the control chart is used to record the history of the sum of the sample measurements, then the control limits are calculated as follows

$$UCL = n\mu + 3\sqrt{n}\ \sigma$$

$$CCL = n\mu$$

$$LCL = n\mu - 3\sqrt{n}\ \sigma$$

Considering another hypothetical example, assume that the diameters of residential plumbing pipes are checked, in groups of five units each, at the end of the fabrication cycle. If the population of pipe diameter measurements is characterized from historical information to have a mean value of 2.010 in. and a standard deviation of 0.010 in. then the control limits are calculated as

$$UCL = 2.010 + \frac{3}{\sqrt{5}}\ (0.010) = 2.023$$

$$CCL = 2.010$$

$$LCL = 2.010 - \frac{3}{\sqrt{5}}\ (0.010) = 1.997$$

Alternatively, using a sample sum control chart gives the following control limits

$$UCL = 5 \times 2.010 + 3\sqrt{5}\ (0.010) = 10.117$$

$$CCL = 5 \times 2.010 = 10.050$$

$$LCL = 5 \times 2.010 - (3\sqrt{5}\ (0.010) = 9.933$$

Figure 4.2 shows a theoretical control chart for the pipe production process where the successive sample averages are plotted with respect to time. From this chart it can be determined that the process is running reasonably well, although there have been periods in which the sample averages have been outside the control limits. The condition of being "out of statistical control" is indicated by a number of control chart characteristics. When one data point falls outside a plus or minus four sigma band or two or more successive points occurs outside of the three sigma control limits then a boundary limit has been exceeded which indicates that corrective action probably should be initiated. Similarly, when seven or more successive points occur on the same side of the mean value, a process shift is likely to have occurred while seven points that "go in the same direction" indicate a trend caused by some sort of process drift. Periodicity in the data also is abnormal from a theoretical point of view, and may indicate a need for a closer examination of the actual situation. Each of these phenomena indicates that something has changed within the manufacturing process and may be cause for concern. Of course, practical judgment must always be utilized in evaluating the severity of a problem. In general, one point outside the three sigma limits is cause for concern and experience may be required to estimate if this is the beginning

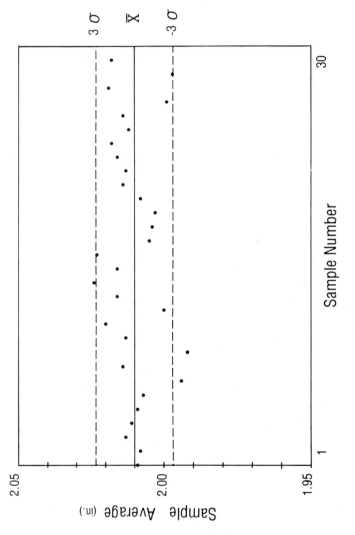

Figure 4.2   Control chart using sample averages (courtesy of E. F. Babelay).

of a problem. If the control limits are close to the process tolerances then these deviations can cause defective products. However, if the control limits are very small with respect to the tolerances then additional leeway exists. Another data pattern that should be recognized is the condition in which the data falls very close to the centerline as opposed to being distributed between the upper and lower control limits. Assuming the data is accurate, this condition indicates that the control limits need to be recalculated so that a meaningful description of the process quality can be obtained. In Figure 4.2 none of the previously mentioned error conditions occur but there was an operating period, depicted by samples seven and nine, during which the level of process control appears to have been marginal.

Figure 4.3 shows another control chart, called a cusum (cumulative sum) plot, in which the deviation of the sample mean from the target value is plotted in a cumulative fashion. Since the sample mean should only differ from the population mean by a random amount with a mean value of zero, the summation of this value should result in a plot that varies about zero with respect to time as long as the process is in control. This chart indicates process shifts by changes in the slope of the data and in this instance it is a more sensitive indicator of process condition than the curve shown in Figure 4.2. Not only does it detect several subtle shifts in the process as indicated by the slope of the data but it also depicts a shift in the sample means that occurred around sample number 23. These charts are most useful in detecting changes of a specific kind since identifying an assignable cause can get quite confusing if more than one change is occurring at the same time.

Other statistics can be presented in a similar fashion on a control chart to provide information about the variation of a manufacturing process. Typical examples include the variance, standard deviation or range of the values that are obtained through the sampling process [1]. If the sample variance or standard deviation is used on a control chart, then the control limits are determined through the use of a chi-square variable (which is not covered in this text). In addition, it is customary to have only an upper control line since lack of variation within the samples is usually not an indication of needed corrective actions. If a less complex procedure for establishing the variability within samples is desired, then the sample range provides a useful alternative. The range chart describes the uniformity that exists between the different members of the sample. This technique provides an estimation of process stability that may not be detected by a characteristic such as the sample average since changes in process variation do not necessarily cause an accompanying change in the average value. (It is desirable that the samples be homogenous which means that the variation within a sample is composed of normal random causes. Then, the variation between sample groups can be attributed to normal random causes as well as special assignable causes.) Assuming, as before, that the process is under statistical control and the population standard deviation is known, then the upper control line for a 3-sigma theoretical control chart is

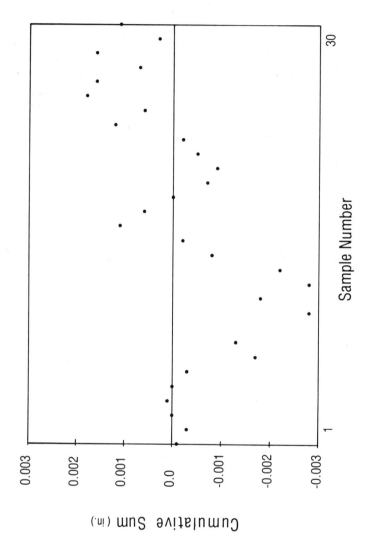

Figure 4.3 Cusum plot for pipe example (courtesy of E. F. Babelay).

$$UCL = D_2\sigma$$

where D_2 is selected from Table 4.2, based on the number of samples in the lot. The lower control limit is usually unimportant for the same reason discussed previously.

Empirical control charts are plots with estimated control lines. They are used in situations in which a manufacturing process is in production but a satisfactory condition of statistical control has not been achieved yet. This could occur because the process is in the initial stages of an operation or it could be due to the existance of one or more major error sources. In either case, the measurements that are taken on successive samples do not exhibit the properties of being a sequence of independent random variables from a stable population. In addition, the goal at this point is frequently to identify the main factors that are preventing the process from behaving in a statistically controlled manner and remove them so that the remaining process variations are characteristic of an in-control process. When this is achieved the magnitude of the process variations within small samples is approximately the same for all the samples, for example, a common distribution exists for the sample measurements.

Estimation of the process mean, $\overline{X}$, for an in-control condition can still be achieved by using the individual means, $\bar{x}(i)$, of the available samples as shown below

$$\overline{X} = \frac{1}{k}[\bar{x}(1) + \bar{x}(2) + \cdots + \bar{x}(k)]$$

Similarly, the estimation of the process standard deviation, $\overline{S}$, for the in-control condition utilizes the pooled value of all the sample standard deviations, $s(i)$,

$$\overline{S} = \frac{1}{k}\Sigma s(i)$$

For sample sizes, n, greater than 25 units, the control limits for sample averages can be determined through the use of the estimated standard deviation.

$$UCL = \overline{X} + \frac{3\overline{S}}{\sqrt{n}}$$

$$CCL = \overline{X}$$

$$LCL = \overline{X} - \frac{3\overline{S}}{\sqrt{n}}$$

For the standard deviation control chart the control limits are calculated in a similar fashion:

$$UCL = \overline{S} + \frac{3\overline{S}}{\sqrt{2(n-1)}}$$

$$CCL = \overline{S}$$

$$\text{LCL} = \overline{S} - \frac{3\overline{S}}{\sqrt{2(n-1)}}$$

For samples of equal size, having 25 or less observed values, the sample mean can be plotted with respect to control limits that are calculated as follows

$$\text{UCL} = \overline{X} + A_3\overline{S}$$

$$\text{CCL} = \overline{X}$$

$$\text{LCL} = \overline{X} - A_3\overline{S}$$

where A_3 is an unbiased estimator for the $(3\sqrt{n})\sigma$ term used earlier in the theoretical control chart calculations. The control limits for standard deviations are

$$\text{UCL} = B_4\overline{S}$$

$$\text{CCL} = \overline{S}$$

$$\text{LCL} = B_3\overline{S}$$

Table 4.3 lists values of A_3, B_3 and B_4 for a variety of sample sizes.

Prior to the proliferation of computers, the calculations required to obtain the value for $\overline{S}$ in the above equations involved extensive manual computations. A widely used alternate approach for estimating the term $(3\sqrt{n})\sigma$ was based on the ranges $r(i)$, of the k samples (as opposed to the standard deviations, $s(i)$, of the samples). If the ranges of the individual samples are assumed to be from normal populations which have equal variances then the average range can be calculated by

$$\overline{R} = \frac{1}{k}[r(1) + r(2) + \cdots + r(k)]$$

Then, an unbiased estimator for $(3\sqrt{n})\sigma$ is $A_2\overline{R}$, where A_2 depends on n and is also shown in Table 4.3. For examples of size 10 or less, this permits the control limits to be defined as shown below

$$\text{UCL} = \overline{X} + A_2\overline{R}$$

$$\text{CCL} = \overline{X}$$

$$\text{LCL} = \overline{X} - A_2\overline{R}$$

The terms A_3s and $A_2\overline{R}$ can be expected to provide reasonably good estimates for $(3\sqrt{n})\sigma$ because they are determined from variations of measurements within samples which are taken over relatively short periods of time and thereby avoid variations due to drifts or shifts with time. (With sample sizes greater than 10 the range statistic becomes rapidly less effective than the standard deviation for detecting assignable causes.) However, it is not appropriate to use $\overline{X}$ as an estimator for the process mean without further modifications. It is a preliminary estimator and should be revised periodically (perhaps every 25 to 50 samples) as

Table 4.3   Control Chart Constants

Sample Size n	A_2	A_3	B_3	B_4	E_2	D_2	D_4
2	1.880	2.659	0	3.267	2.660	3.686	3.267
3	1.023	1.954	0	2.568	1.772	4.358	2.575
4	0.729	1.628	0	2.266	1.457	4.698	2.282
5	0.577	1.427	0	2.089	1.290	4.918	2.115
6	0.483	1.287	0.030	1.970	1.184	5.078	2.004
7	0.419	1.182	0.118	1.882	1.109	5.203	1.924
8	0.373	1.099	0.185	1.815	1.054	5.307	1.864
9	0.337	1.032	0.239	1.761	1.010	5.394	1.816
10	0.308	0.975	0.284	1.716	0.975	5.469	1.777

Source: Ref. 1 & 3.

the process moves toward the desired state of statistical control. In addition, during the time in which a process is being improved it is not appropriate to include data points in the control limit calculations which are correlated to specific process conditions that have been corrected. However, a data point should not be discarded just because it falls outside of a control limit. The UCL for ranges is calculated, using the appropriate value of D_4 from Table 4.3, as

$$UCL = D_4\overline{R}$$

The previous paragraphs have discussed situations in which data are selected as samples of size n out of larger populations of size N and used to generate a statistic that is plotted on the control chart. This averaging effect is useful in smoothing out the "noise" variations in the data. However, there may be other instances in which it is useful to plot each individual data point. Examples of this situation exist when it is desirable to obtain feedback information after each part is fabricated or after each step of a process. Obtaining sampled data is not appropriate in this instance because of the potential impact of the delay in receiving the information. If the value of the product is relatively high and the feedback information is needed to monitor control actions that are being implemented, then waiting until a complete group of items have been fabricated and tested is unacceptable. The use of control charts for individuals also is appropriate when the sampling process is costly or destructive or when the sampled material is quite homogenous.

If there is a logical rational such as time of day, shift, and so on, for grouping these individual data points into batches, then the equations shown below can be used to calculate the control limits that are used

$$UCL = \overline{X} + E_2\overline{R}$$

$$CCL = \overline{X}$$

$$LCL = \overline{X} - E_2\overline{R}$$

for plotting the individual data points. The constant E_2 is shown in Table 4.3 for different sizes of groups. If logical groups do not exist then the same equations can still be employed, but the variation (moving range) between points must be used to generate the $r(i)$ values used to calculate the control lines. If a moving range of size 2 (difference in adjacent points) is utilized then the constant E_2 that is obtained from Table 4.3 for the control limit calculation is 2.66.

The previous control chart discussions have dealt with process characteristics that can be given a numerical value. An alternate approach is to utilize the attributes information discussed previously. In this case a factor p is designated as a "fractional defective" and used to describe a process on a go/no go basis. This factor represents the number of the occurrences that meet the conditions of interest divided by the entire number of items being considered for a particular sample. In most cases, this technique is useful for large samples (greater than 50 to 100 units) which have 4% or more defective units [3].

Assume that the total number of units tested are divided into equal subsets, and the average defective, $\overline{p}$, is defined as the total number of defectives in all the samples divided by the total number of units in all samples. Then, a control chart can be used to analyze the output of a process by plotting the number of defects in each sample in a sequential fashion. In a similar manner to the previous control limit equations, the central control line is $\overline{p}$, and the control limits are

$$\overline{p} \pm 3 \left[\frac{\overline{p}(1 - \overline{p})}{n} \right]^{\frac{1}{2}}$$

For values of $\overline{p}$ less than 0.1 these limits become

$$\overline{p} \pm 3 \left[\frac{\overline{p}}{n} \right]^{\frac{1}{2}}$$

An optional approach is to utilize a control chart for plotting pn, the fractional defective times the sample size. In this case, the central control line is equal to $\overline{p}n$, and the control limits are

$$\overline{p}n \pm 3 \left[\overline{p}n(1 - \overline{p}) \right]^{\frac{1}{2}}$$

where $\overline{p}n$ is equal to the total number of defectives in all samples divided by the number of samples (average number of defectives in the individual samples). Again, when $\overline{p}$ is less than 0.1 the control limits can be simplified to

$$\bar{p}n \pm 3\,[\bar{p}n]^{\frac{1}{2}}$$

which is plus or minus three times the square root of the average number of defectives in samples of equal size.

Similar equations can be used to prepare control charts for the number of defects per unit and the number of defects. If $\bar{u}$ is considered to be the average number of defects per unit in the complete set of tests, then it can be calculated as the total number of defects in all samples divided by the total number of units in all samples. Then, the central control line is $\bar{u}$ and the control limits are

$$\bar{u} \pm 3\left[\frac{\bar{u}}{n}\right]^{\frac{1}{2}}$$

If the parameter of interest is the number of defects per sample, $\bar{c}$, then the central control line becomes the total number of defects in all samples divided by the total number of samples. The control limits are

$$\bar{c} \pm 3\,[\bar{c}]^{\frac{1}{2}}$$

One example of the benefits of employing control chart techniques as a part of a quality control effort is demonstrated by the experience of a cutting tool manufacturer [4]. New York Twist Drill is a manufacturer of tight-tolerance twist drill bits for the aerospace, automotive and general metal working industries. Prior to the serious application of statistical process control methods, New York Twist Drill was experiencing scrap rates of about 13% and a rework rate of about 15%. Following a top-down commitment to the statistical process control philosophy, quality control techniques were implemented that were based upon in-shop control charts. These changes resulted in the machine operator's having a sense of ownership and increased pride in the manufacturing processes which provided almost immediate cost savings. Scrap was reduced to less than 1% and rework costs were reduced by about 75%. In addition, each step in the manufacturing process became easier to accomplish due to the cumulative benefit of having improved consistency and quality occurring throughout each of the preceding stages.

Error Budgets

Error budgets are used to predict and control the total error of a manufacturing process. The assumption is made that the total error is composed of a number of individual error components that combine in a predictable manner to create the total system error. (This is the same assumption that was introduced in the earlier statistics chapter when the subject of probability distributions was discussed, and is a continuing assumption throughout this text). The task is to identify each of the error components and isolate the physical causes so that they can be characterized and controlled or monitored. In particular, it is necessary to know which components are the major causes of variability in the process. Then, corrective action can be taken to reduce the sensitivity of the process to these error sources or at least monitor them so that the excursions are detected before it is too late.

If a linear function of random independent variables exists such that the overall error, E, is a function of the individual error components, $e(i)$, then

$$E = c(0) + c(1)e(1) + c(2)e(2) + \cdots + c(n)e(n)$$

where the $c(i)$ terms are constants. Since the $e(i)$ terms are uncorrelated, the variance may be expressed as a combination of the individual variances:

$$V(E) = c(1)^2 V[e(1)] + c(2)^2 V[e(2)] + \cdots + c(n)^2 V[e(n)]$$

While the magnitude of the variance of the individual error components is a factor in determining the major error source in a given situation, the constants, or sensitivity factors are also very important. A particular error factor may exhibit a lot of variation but its effect on the process can be negligible because the process is insensitive to these variations. An example is the attempt to perform a high accuracy measurement on an aluminum part in an area that is subjected to wide temperature excursions. If the measurement of interest is the surface texture of the workpiece, the temperature variations are unlikely to influence this characteristic (although it may influence the gaging system). However, if size measurements are attempted, then this feature is likely to be perturbed by the temperature excursions.

If the total error is a nonlinear function of the individual errors then the problem is more complex unless linearity can be assumed to apply over a restricted area of operation. One way to deal with this problem is to calculate the value of the particular function at the point of interest and at points that are three sigma above and below this point. If the slopes of the curve above and below the base point are similar over this range, then linearity can be assumed. In addition, the value of the slope gives the gradient or sensitivity of the function at the particular operating point.

In most manufacturing situations, the procedure for obtaining a viable error budget begins with the identification and characterization of the significant error sources. Then a combinatorial rule must be selected for generating a total error from these individual error components. In many situations, an averaging process occurs so that a statistical approach may be taken to obtain the total system error. One useful technique in this case, is to utilize the root mean square or RMS values of the individual error sources. Additional information on using this technique for process control will be discussed in Chapter 6, which covers process monitoring.

An excellent example of the use of this approach for making a judgment trade-off during the design of a complex system is the work that was performed at Lawrence Livermore National Labortory for the Large Optics Diamond Turning Machine (LODTM) [5]. Based on the need for reflective metal optics of greater size and accuracy than were previously available, the Department of Defense and Department of Energy cooperated to develop a preliminary design for a state-of-the-art diamond turning machine. This machine was designed to be capable of manufacturing workpieces up to 60 inches in diameter and 12 inches in length, with weights as high as 2,200 pounds, with a figure error tolerance of 0.5 microinches. Figure 4.4 shows an artist's concept of the machine design that was originally proposed. In this situation, the machine output is optical components which are more sensitive to averaged errors that they are to localized spot defects of similar magnitude. This means that the RMS error value is more meaningful that the amplitude of the worst isolated defect. The combinatorial rule chosen for obtaining the total estimated RMS error caused by a number N of independently varying error components was

$$\mathrm{RMS}_{tot} = \left[\sum (\mathrm{RMS}_i)^2 \right]^{\frac{1}{2}}$$

where RMS_i is the RMS amplitude of the individual error elements.

In this instance, the accuracy goals exceeded the state-of-the-art for diamond machining with a machine of this size. The error budget technique became a

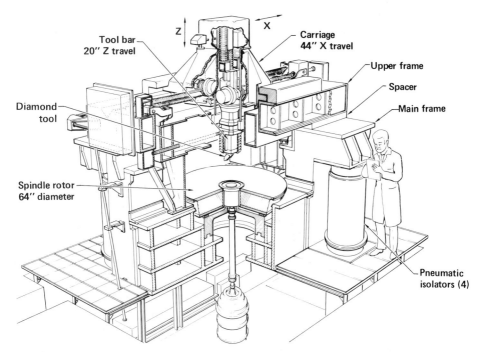

Figure 4.4   Artist's concept of initial LODTM design (courtesy of Lawrence Livermore National Laboratory).

valuable tool in evaluating the impact of the various error sources on potential design options and for selecting the most advantageous areas for attempting to extend the state-of-the-art.

In general, when working with error budgets, it is desirable to attempt to reduce the amplitudes of those error sources that can be made relatively small with only a modest effort. However, there is certainly a point of diminishing returns beyond which the expended resources are essentially wasted. In addition, it is frequently the larger error sources that require the most effort to achieve a suitable reduction in amplitude. As shown in Chapter 3, once an error has been reduced to 10% of the dominant error, there is little incentive to attempt to reduce it further because the dominant error overrides the other factors.

Another way of utilizing an error budget is to consider it as a modeling technique for describing the performance of a process. As such, the individual components of the model, or error budget, are subject to verification through experimental measurements just as the total model may be tested for accuracy. In most cases the final model is derived through a process of iteration in which various error factors and methods of combining these error factors are evaluated for suitability. An offshoot of this sequence is that an improved understanding of the process is obtained. This can lead to appropriate process optimization in many instances. Evaluation of the final error budget, or model testing, is necessary in that it verifies the assumptions that have been used in the modeling process. This means that the model should be tested under a variety of operating conditions. This will allow the determination of whether or not the process output is accurately predicted under the range of operating conditions that are likely to occur. If the model should prove to be inadequate this provides information that is used to achieve an improved understanding of the process.

References

1. I. Guttman and S. S. Wilks, *Introductory Engineering Statistics*, John Wiley & Sons, Inc., New York, pp. 126–128, (1967).
2. M. R. Spiegel, *Theory and Problems of Statistics*, Schaum Publishing Co., New York, p. 157, (1961).
3. *Presentation of Data and Control Chart Analysis*, American Society for Testing and Materials (ASTM), ASTM Special Technical Publication, Philadelphia, Pennsylvania, (1976).
4. SPC charts progress, *Automation*, *35*:1 (1988).
5. R. R. Donaldson, Large optics diamond turning machine, *Volume I, Final Report*, Lawrence Livermore National Laboratory, UCRL-52812, Livermore, CA, (1979).

CHAPTER 5

Sensors

Introduction

In the manufacturing environment, sensors are frequently involved in the general areas of qualification, control and monitoring. [1] Qualification is a procedural function, where the results of an operation are inspected to be sure that they meet the necessary requirements or tolerances. Qualification activities usually bring about monitoring guidelines or control requirements which direct some action that is taken as a result of the data obtained in the inspection process. Some typical qualification tasks include leak detection, dimensional measurements, determination of mechanical or electrical properties, alignment checking, go/no go functionality and color differentiation.

Controlling the quality of manufacturing processes is a wide spread goal in the competitive market place. However, before the level of quality can be controlled, it is necessary to be able to measure it. This is accomplished by sensing accurately those particular characteristics that interact to determine the quality of an operation or product. Since there is rarely any rational justification for the production of low quality products, it is usually desirable to seek a high quality level of manufacturing. This does not mean that it is necessary to build a system that never makes a defective workpiece, since this may be inappropriate due to the costs that would be incurred. However, it is generally useful to be able to characterize the process quality while production is underway. The alternative to this approach is to wait until problems become evident through the production of faulty products, but this is usually not an attractive mode of operation. If the quality features of the product cannot be measured while the process is under-way, then secondary features must be substituted to estimate the condition of the process output.

Sensors are a key element with whatever approach is taken for monitoring manufacturing operations. They provide the information feedback from the factory floor or the inspection area that permits an estimation of whether or not the quality of the process output meets the design specifications. In some situations, sensors are utilized in an open-loop fashion to provide an operator with real-time information that defines the status of some feature of the process. In other in-stances, the sensors are an integral part of the manufacturing system and they are used to provide information to a feedback control loop that determines the manner in which the operations occur.

Sensors are usually involved in three generic types of monitoring applications. Production monitoring is one use for sensors in which the transducers are uti-lized to determine the status of operations on the production floor. One frequent use for these sensors is to answer questions concerning the amount of material left which requires processing, the total number of parts produced, the number of good or bad parts, up-time, down-time, cycle time, and so on. Machine monitor-ing is a second application for sensors which has been discussed in previous chap-ters. This activity involves the determination of whether or not a process is functioning properly. An early warning of the need for preventive maintenance or process adjustments is the objective of these measurements. The third applica-tion area, environmental monitoring, provides information concerning the con-dition of an area. A common location for the installation of these sensors is in the heating, ventilation and air conditioning system.

In general terms, performance requirements for sensors involve the need to detect the presence/absence, position, condition or identity of an object. As might be expected, each of these broad catagories overlaps with the others to a certain extent. This is because one particular transducer may be applicable for use in multiple areas depending upon the manner in which it is employed. However, this classification still provides a useful guideline for discussion purposes.

Sensors that are utilized to detect the presence or absence of an object are gener-ally less complex than the other types mentioned above because it is only neces-sary to detect whether or not an object is occupying a particular point in space. The measurement system is designed only to detect when the position of interest is occupied not where an object is when it is not at this location. These devices usually only output a digital or on-off signal because there are only two condi-tions of interest in this situation. Some potential sensors that might be used for this type application include photoelectric units, mechanical or electrical limit switches and proximity switches.

Position sensors may be described as a refinement of the presence/absence sensors. In this case, the measurement capability of the transducer is enhanced to the point at which it is possible to determine where an object is located, how far it has moved, and how close it is to something else. These devices are a basic element of position control systems and as such are used throughout the manufacturing process. Some of the sensors that may be used to perform the position measurement function include rotary or linear encoders, resolvers, electronic scales and laser interferometers.

The condition sensor category can be quite large since these devices can be used to inspect a workpiece or to measure other parameters associated with a manufacturing operation. These devices are able to provide information about the status of an entity at a particular point in a manufacturing cycle. Examples of some of the workpiece characteristics that might be measured to provide product inspection information include size, shape, mass, concentricity, flatness, chemical composition, color and so on. Additional condition sensors might be utilized to provide process information such as electric current, vibration levels, flow rates, pressure, force, temperature, humidity, and so on. In addition, devices such as the laser interferometer may find applications in both the condition and position categories. This is demonstrated by the noncontact inspection gage [2] shown in Figure 5.1. In this example, the laser beam is split into two paths: one path measures the position of the inspection machine's linear axis, while the other beam is focused onto the part surface to measure roundness and surface finish (condition).

Identification sensors are used to differentiate between different types of parts as well as to measure the identity or serial number of a specific workpiece, pallet or other item. The range of requirements for these devices can vary from just sorting the "big ones from the little ones" to reading information from a bar code label or an "imbedded" semiconductor. Photoelectric sensors can readily sort parts based on height differences but they are not sophisticated enough to know which part is present (unless size is the only criteria for determining this characteristic). Recently, visiontype sensors have received a lot of attention for these particular applications since they can simulate human visual inspection activities. In fact, in some cases the machine-vision system can surpass the performance of a human in terms of speed and resolution.

Sensors are essentially analog in nature in that some physical characteristic is measured in the sensing process. Even when the sensor is measuring the presence or absence of something such as reflected light, an electric current or a magnetic field there is still a specific level that the sensed property must exceed before the status change is detected. However, the definition of a sensor as being digital or analog does not refer to this characteristic. This differentiation involves the manner in which the sensor system processes the measured information. A digital sensor such as an optical encoder produces counts that correspond to incremental changes in the sensed parameter (rotation), and an external counter provides

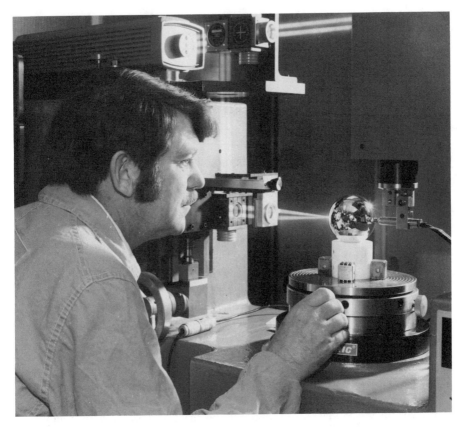

Figure 5.1   Noncontact laser inspection machine (courtesy of Martin Marietta Energy Systems, Inc.).

a running tally of the distance that has accumulated. The resolution limit of the system is a function of the sensor (the divisions on the encoder disc), but the range is limited only by the size of the external counter and is essentially unconstrained by anything except the ability to count to a specific value. In contrast, a purely analog sensor such as a pressure gage is limited to a specific range. The term digital-output sensor refers to an analog sensor which has the necessary internal electronic circuitry to provide digital output information. This type of device can be interfaced to equipment such as a digital computer, but it is still limited to a specific range that is determined by the physical limitations of the specific transducer.

Sensors which are used for analog measurements are affected by a range/resolution limit that does not apply to digital transducers. This means that as the resolution of the analog device is increased the useful operating range is also decreased. An example of this constraint is easily demonstrated using a mercury thermometer.

A thermometer used for cooking may have a range of over 100 degrees centigrade with a resolution of 1 degree. In contrast, a thermometer used to measure body temperature has a resolution of a fraction of a degree but a range of only a few degrees. The body temperature thermometer could be built with an extended range but the device would be too long to be used safely. One solution to this problem is to utilize a different sensing system such as a thermocouple with a digital readout. This improves the immediate situation but there still is a range/resolution limit that is determined by the physical characteristics of the thermocouple. In a similar fashion, the accuracy of both analog and digital measurement transducers is degraded as the full-scale range capability of the devices is extended. However, an advantage of a digital system (which may consist of an analog sensor and a digital computer) is that the device can be calibrated readily. This means that the errors can be automatically corrected so that the useful range is extended.

The connection between a particular sensor and the "outside world" is called an *interface*. This interface provides the collection point at which the sensor-based information is entered into a monitoring or controlling system. Sensor interfaces are able to process inputs from a wide variety of transducers and send this data on to programmable controllers, computers or other interface devices. In some applications, a local system decision is made based on sensor data and no interaction is required with other portions of a system. An example of this type of action is a closed-loop position servo system that is associated with a solitary machine control unit. In this situation, position feedback data is sent to a local collection point that initiates the proper response to this information. In the event of a malfunction, a local alarm is generated but no external communication is required with any remote systems. In contrast, if the same system is connected to a host computer, then it may be desirable to signal the alarm condition to the host computer in order to initiate the necessary maintenance activities or just to maintain a record of the system status.

This chapter is concerned with the various types of sensors or transducers that are available to monitor the operating conditions within a manufacturing process. The general division of position/condition sensing will be utilized for convenience in organizing the material within the text. However, no other significance should be attributed to this particular grouping. Sensors for detecting the identity or presence/absence of an entity are considered to be a subset of condition sensors for organizational purposes in the remainder of this chapter. A brief description is included of a variety of sensors that are available for measuring the different characteristics of manufacturing operations, as well as the interfaces that are available to connect the sensor information to a data processing unit. Supporting material is provided on a topic called the *Sampling Theorem* which describes the frequency with which the sensor output must be monitored in order to capture the dynamic properties of a particular signal. In addition, examples of sensor applications in a variety of maufacturing conditions are included and special attention is given to machine vision systems. Machine vision is a relatively new technology and a description of the integration of these sensors with the peripheral equipment that is necessary to obtain a complete operating system provides a good overview of the integration of sensors into an automated manufacturing environment.

Position Sensors

This group of transducers is utilized to measure the dimensional characteristics of an object, as opposed to other features, such as color, tensile strength, electrical conductivity, etc. These measurements are utilized to generate the information required to define the dimensional nature of the object as well as it's displacement with respect to a reference location. Common applications include measuring workpiece size and shape, detecting when an object is in the desired location for an operation to be performed, providing position feedback information for machine drive systems (where the sensed displacement is either a rotary or linear motion), or the measurement of machine error motions (such as slide straightness errors).

Rotary Encoders.   These rotary transducers are used to detect angular motion through the use of discs which have an encoded pattern that is precisely located around the disc's periphery. Angular displacement is detected by a measurement transducer that is able to infer relative rotational motion between the disc and the transducer. An optical encoder utilizes a disc that has alternating opaque and transparent sections as shown in Figure 5.2. Light passing through the disc to photodetectors is interrupted by the opaque sections so that a series of pulses is generated by the rotary motion of the encoder shaft. Absolute encoders provide an output in the form of a digital word that describes a unique angular position. Magnetic encoders use a magnetic encoder wheel to store the

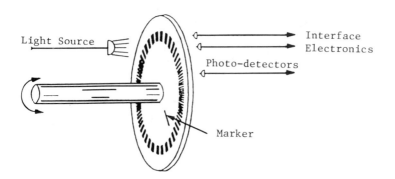

Figure 5.2   Optical-encoder disc attached to rotating shaft.

encoded information that is decoded to infer displacement. Fiber optic encoders employ a rotary code wheel that has coded, and high and low reflectivity sections that extend around the cylindrical periphery of the disc. The presence or absence of the reflected light provides the signal that is converted into angular displacement information. Incremental encoders employ a pulse output that is indicative of relative motion but requires external circuitry to maintain a relationship to an absolute position. However, absolute encoders provide an absolute indication of angular position that does not depend upon external circuitry and these units are unaffected by loss of power to the system since the absolute position is available from the encoded signal at all times.

Encoders are frequently attached directly to a motor shaft although gearing may also be used to increase the system's resolution. Typical numbers of divisions per revolution range from 50 to 18,000. Higher resolution units are available with over 1,000,000 pulses per revolution through the use of resolution extension techniques. A frequent use for encoders is to infer linear motion in rotary drive systems. For example, with the machine slide shown in Figure 5.3, the assumption is that the rotary motion of the position transducer shaft is directly proportional to the linear motion of the machine axis. While this is rarely the case, due to errors in the mechanical gearing as well as the geometry of the machine's slides, it is frequently a suitable approximation.

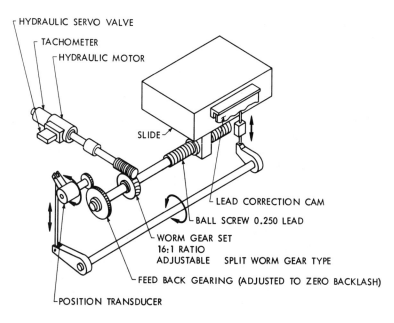

Figure 5.3 Sketch of machine tool slide drive system (courtesy of Martin Marietta Energy Systems, Inc.).

Resolvers. These transducers are similar to AC electric motors in that they are electromechanical devices. These sensors have two stator windings that are physically located 90 degrees to each other. Another winding is attached to the device's rotor. Electrical excitation of the unit with an alternating current coupled with motion between the rotor and stator produces an electrical output signal that defines the degree of rotary motion. These rotary sensors are used in many of the same applications as optical encoders. However, because of their relatively coarse resolution (with respect to an encoder), they are frequently used with mechanical gearing, as shown in Figure 5.3, to obtain the desired positioning resolution. (This figure also shows one technique for utilizing a leadscrew correction-cam to mechanically correct axial positioning errors on a machine slide. This is accomplished by modifying the position of the resolver body with a compensation-cam bar linkage. This alteration in the feedback information induces an adjustment in the slide location which corrects the previously calibrated positioning errors.) External electronic circuitry is usually required to excite these devices and to process the output signal, although some units exist that are as simple to use as an encoder. A typical pulse resolution value for a resolver system is 4,000 counts per revolution. For the system shown in Figure 5.3, with a 0.250 pitch leadscrew and 6.25:1 feedback gearing, this would provide a linear resolution of 10 microinches.

Linear Scales. These devices use either electrical, magnetic or photoelectric sensors to convert the linear motion between the moving and stationary members of the transducer into an electrical output signal. These devices can be thought of as linearized encoders or resolvers that can be used to operate a position display or to provide feedback information to a closed-loop control system. Since these systems provide a direct measurement of linear motion there is no need for the mechanical gearing shown in Figure 5.3. Although this lack of gearing also eliminates one possible technique for extending the positioning resolution of these systems, a significant advantage is that linear motion is measured directly rather than being inferred from angular motion. Also, these units are relatively rugged and can be imbedded within the body of a machine when necessary. Measurement resolutions are available from below 10 microinches to over 0.1 in. Useful measurement ranges generally vary inversely with resolution and can extend from a few inches to about 100 ft. Scale accuracy is usually stated as a certain basic value plus a factor that depends on the length of the scale. However, the accuracy of the installed position measurement system is dependent on other factors such as the alignment errors between the machine axis and the measurement axis of the transducer. In addition, the inherent positioning errors due to the angular motions of the machine axis (roll, pitch, yaw) can further degrade the measurement accuracy of the system because the lateral offset between the physical position that is being monitored and the axis of the measurement transducer. When appropriate, system accuracy can be enhanced by mapping these and other system errors that are related to the positions of the machine components.

Then, this calibration information is used to correct for the machine's repeatable positioning errors.

Laser Interferometer. Currently, the most widely used light source for optical interferometer displacement measurements in manufacturing is the helium neon laser. Through the use of stationary and moving optical elements attached to machine members, the laser measurement system is able to detect relative motion with a resolution of 0.1 microinch or less, with an accuracy of one part per million and a range of over 100 ft. In addition, fringe interpolation systems are available for use with interferometers that divide the basic light wavelength by a factor of 2,000 which produces a resolution of 0.013 microinch. (Of course vibration can be the limiting factor at these levels.) While it is probably the most accurate position transducer commercially available [3], it is also particularly susceptible to environmental disturbances. When the laser beam is operating in the open atmosphere, allowances must be made for the temperature, pressure, and humidity of the air. In addition, the need for unobstructed laser-beam pathways is a necessary design constraint, although laser tracking systems exist for use with robots or other random motion mechanisms that are not limited to the motions allowed by a specific set of guideways. This type of device depends on a tracking system to follow the motion of the object that is being observed. In contrast, Figure 5.4 shows the laser head and the associated optical elements installed on a machine tool with independent slide ways. The optics are all covered with plexiglass shields to protect the system from chips, coolant or other objects that would interrupt the beam path. Since the accuracy of this displacement transducer is also influenced by angular slide motion errors it is desirable to mount the optics as close as possible to the point of interest. However, there are frequently physical constraints that prohibit this, just as happens with other linear displacement transducers. Nevertheless, for those applications requiring the ultimate in measurement accuracy, the laser interferometer is unsurpassed. Unfortunately, it is also the most expensive displacement measuring system that is routinely available for use in making precision position measurements over an extended range.

Reflected Light Transducers. These noncontact measurement systems function through the use of a light beam that is reflected off of the target surface. The reflected beam strikes a photodetector array at different locations as displacement occurs. This translation of the light across the detector array surface is used to sense relative motion between the probe head and the target surface. The resolution of the optical triangulation systems varies from below 100 microinches to about 500 microinches and is limited by the reflection angle and the coarseness of the detector array. Standoff values and measurement ranges are available from a fraction of an inch to several inches so that the useful applications areas are different from those generally served by laser interferometers or linear scales.

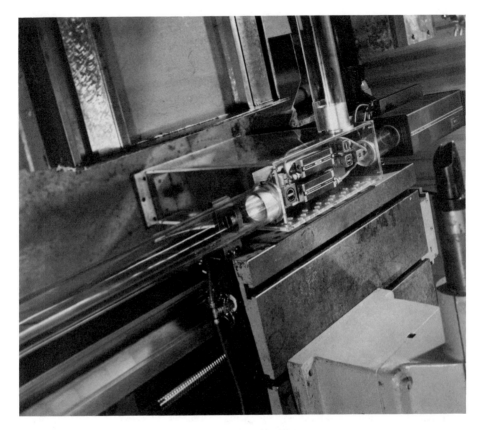

Figure 5.4   Laser interferometer displacement transducer mounted on a machine tool (courtesy of Martin Marietta Energy Systems, Inc.).

These noncontact systems have a more limited frequency range than a capacitance gage and they only work with reflective surfaces. However, they have a definite advantage in some applications because they are relatively inexpensive and they are available as noncontact inspection probes for coordinate measurement machines. For extremely short ranges of displacement requiring a high measurement accuracy it is also possible to utilize the difference in the intensity of reflected light from a reference and target surface to obtain a measurement resolution of 0.04 microinches [4].

Hall Effect Sensors.   These devices are based on semiconductors that respond to a magnetic field. One characteristic of these sensors is that when a magnet comes near a Hall effect device, through which a current is flowing, a voltage is induced. Also, it is possible to utilize a Hall effect sensor that has an integral magnetic field that responds to changes that are induced in this field. The units must be sealed in a package which protects it from external contaminants. These sensors are suitable for use in applications such as limit switches, encoders, tachometers, and so on.

Linear Variable Differential Transformer (LVDT).

These units are short range contact transducers that have good stability, low noise and low drift but they are also very nonlinear over-extended ranges. The typical approach is to operate these devices only over their linear range, which decreases with increasing resolution, but calibration techniques exist [5] that use a microcomputer to linearize the device output over an extended range of travel. The quill of the inspection machine shown in Figure 5.5 is supporting an air bearing LVDT that has a readout resolution of 10 microinches, a range of travel of 0.020 in. and a gaging force of a few ounces. The LVDT is used in this application to inspect a cutting tool for "bestfit" size and shape. The center of the spherical cutting tool is aligned to the rotary table center of rotation in order to correctly position the tool edge prior to the inspection process. Then the edge of the cutting tool is swept by the LVDT as the rotary table is rotated. Deflection of the probe produces an analog signal that is sent to a chart recorder to provide a graphic output of the part roundness or surface finish. In this application the axis of the transducer is perpendicular to the measurement surface. However, because the air bearing LVDT has such a low gaging force, it is possible to make measurements on surfaces that are inclined as much as 45 degrees to the axis of the probe, as long as the output data is corrected appropriately. Other applications for LVDT's include use as position null devices in systems designed to provide datum locations and as measurement probes on coordinate measuring machines.

Capacitance Gage.

Capacitance sensors consist of two or more conductive surfaces which are separated by a dielectric media (nonconductor). The value of capacitance is varied by changing the distance between the plates, the size of the plates or the dielectric constant of the separation media. It is relatively easy to use capacitance sensors for the measurement of object proximity or displacement. As a proximity switch, the capacitive sensor can measure the level of a variety of materials by looking through the side of a container (assuming a nonmetallic container) as well as perform the more common material presence/absence applications. For displacement measurements, the capacitance probe is similar in size, range and resolution to an LVDT. In either of these application categories, the noncontact gage measures the capacitance between the probe and a target object which is a function of the spacing between the probe tip and the target surface. Therefore, the transducer is able to interpret changes in capacitance as a relative motion between the probe and measurement surface. (For special

Figure 5.5 Air bearing LVDT being used to inspect size and shape of diamond-tipped cutting tool (courtesy of Martin Marietta Energy Systems, Inc.).

applications, capacitance gages can also be used to measure surface roughness per ANSI specification B46.1. [6]) The only requirements are that the target material be electrically conductive, at the carrier frequency used by the system, and that the material (usually air) in the gap between the probe and the part surface has a constant dielectric constant. Because it is a noncontact analog transducer and inherently has a relatively high useful frequency range (100,000 hz), it is especially useful for detecting things like spindle motion errors at high speeds [7]. Other applications include a gravity tilt sensor in which a dielectric is used to cover an increasing amount of the plate area as the tilt angle changes and an acceleration sensor that is fabricated from a single crystal of silicon.

Capacitance probe sensors are available in a variety of sizes and shapes for special applications. One operating constraint is that it is influenced by surface contamination so that surface cleanliness is required if meaningful results are to be obtained. In addition, the transducer inherently averages the characteristics over a specific area so that spurious results can be encountered near workpiece edges. Also, the standoff range typically becomes smaller as the resolution is increased so that care must be exercised in these gaging operations to insure that neither the probe nor the workpiece surface is damaged through accidental contact. While it has a wider frequency range than an LVDT and offers the advantages associated with contact gaging, it is also usually more expensive.

Eddy-Current Transducers.

These noncontact measurement systems detect the impedance that is encountered by eddy currents that are induced in a conductive metal target material. As the distance between the probe and the target changes, the impedance also changes. This feature provides an electrical characteristic that can be converted into a measure of displacement. Since the system is also sensitive to the electrical characteristics of the target material it is necessary to calibrate it for different metals. However, it is insensitive to a nonconductive medium between the sensor and the target so that contaminants such as oil, dirt and moisture do not degrade the system accuracy. Typical resolution values vary from 0.0001-0.0005 in. for measuring ranges from 0.020-2.0 in. with a temperature sensitivity of 40 microinches to 0.002 in. per degree Fahrenheit over the resolution range. These transducers produce a DC output signal that is proportional to the average gap between the probe and the measured surface as well as an AC component that represents the dynamic motion component. A frequency range of 0-10 KHZ is readily available.

Reluctance Transducers.

This is another electric-field-effect system similar to the capacitance and eddy current transducers. Its measurement capabilities resemble those of the eddy current systems although it is limited to use in ferromagnetic materials (it can also be used to detect the presence of ferrous material). As with the eddy current transducers, increases in measuring range require increases in sensor size. This also increases the field of view so that the measurements are averaged over a larger area.

Precision Potentiometers. These transducers are similar to encoders or linear scales except that the output is an analog signal rather than a digital signal. These variable resistors are available in either rotary or linear construction and may be obtained with a brushless design if needed. Also, a variety of mathematical transfer functions can be specified. Typical linearity values are better than 0.1% of full scale, but this value is relatively coarse for displacements of an inch or more, when compared to previously mentioned transducers.

Ultrasonic Sensors. These devices depend on the reflection of sound waves from a target to determine position either as an absolute measurement or as a detection of proximity. This technique is similar to that used by bats, automatic door opening systems and self-focusing cameras. Assuming the speed of the sound waves is constant throughout the medium of interest, then it is possible to calculate distance by measuring the time for an acoustic energy burst to travel from the transmitter to the target surface and back to a receiver (although a calibration measurement on a known thickness sample may be required). This reflection time is proportional to the thickness of the material and is influenced by the mechanical properties of the object being measured. In situations where the material to be sensed has varying mechanical properties from one part to the next, it is possible to utilize a self-calibrating technique that compensates for the changing conditions, although local variations in the speed of sound will still cause accuracy problems. In addition, it is possible to measure gaps and voids by sensing the multiple reflections that occur from the different surfaces. Typical resolution values are in the range of 0.0001 to 0.001 in. Depending on the application, these sensors generally are based on one of three different types of transducer: piezoelectric, magnetostrictive or electrostatic devices. Piezoelectric ceramics commonly used for ultrasonic measurements include single crystal-ADP, barium titanate, lead zirconate titanate, lead metaniobate, lead titanate, and lead magnesium nibate. One example of a magnetostrictive device consists of a stationary ultrasonic waveguide and a moving permanent magnet. When a current pulse is sent along the waveguide it sets up a magnetic field circumferentially over the entire length of the waveguide that reacts with the magnetic field of the moving permanent magnet. This phenomena allows the location of the moving magnet to be detected with respect to the ultrasonic driver. Using different data manipulation techniques resolutions are available ranging fom 0.001 in. down to microinches [8]. One advantage of this system is the ruggedness of the device, another is the ability to configure the waveguides in the form of a flexible rod so that measurements of curved motion can be accomplished. Applications vary from the detection of liquid levels in tanks to position measurements on injection molding machines, hydraulic cylinders and wavemaking machines.

Pneumatic Sensors.   One use for these gages is to employ differences in air pressure to determine the distance to a target object or the size of a hole. Following a calibration measurement on a master part, the air jet is directed at the surface of interest and the back pressure resulting from the air flow restriction is used to detect the proximity of the object. These sensors have good resolution, although they are limited to a relatively small measurement range. Therefore, they frequently are useful as go/no go gages in those instances where it is necessary to measure the average characteristics of a surface or hole. Another use for pneumatic sensors is measuring the ability of a workpiece to maintain a specific level of vacuum or pressure over a period of time (leak testing).

Vision Sensors.   These devices are used in systems that range from a complex camera/computer package that identifies types of objects and determines their location and orientation to simple light beams and photo detectors that sense the location of an object in the system's field of view. A matrix-array vision system evaluates the projection of an image onto a two dimensional photo detector surface. Resolution is dependent on the number of sensing elements (pixels) in the photo detector array, although lenses may be used to increase the system magnification at the expense of the field of view. Also, interpolation techniques are available to provide sub pixel resolution. A 1024 × 1024 array provides a resolution of approximately 0.001 times the field of view. Limits to measurement accuracy include system linearity, resolution, and stability. A significant advantage is the speed with which the noncontact measurements can be made, a disadvantage can be the limited field of view at higher resolutions.

Touch-Trigger Probes.   These transducers contain a sensing device that is activated when the probe stylus touches the measurement surface. Some probes require a slight deflection of the probe stylus to generate the contact signal while others require only contact with the surface being inspected. Because the probes are very repeatable and omnidirectional they are widely used on coordinate inspection machines. System resolution is limited primarily by the resolution of the machine axis that is used to position the probe. Accuracy is limited by the measuring machine's inherent accuracy and the care that is taken to master the probe for a particular application. Since some of the units utilize a probe with a mechanical switch, there is a small but finite deadband. In addition, the trigger characteristics may not be identical in all directions and because the size and shape of the stylus tip may be unknown, it is important to calibrate the system for the types of measurements that are desired. However, if care is taken in the measurement process it is possible to obtain meaningful information with a resolution of 10 microinches or less.

Figure 5.6 shows two types of tough-trigger probes that can be mounted on the same tooling that is used to hold cutting tools. Two electrical contacts are available on the shank of the tool holders that are dedicated for use with the gaging

Figure 5.6 Tool holders for use with cutting tools and wireless touch probes (courtesy of Martin Marietta Energy Systems, Inc.).

probes. This allows the probe output signal to be transmitted to the interface electronics via mating contacts in the tool holder receptical so that the units are compatible with automatic tool changers. If a hardwired connection is not a problem then a mechanical clamp such as the one shown in Figure 5.7 may be an acceptable approach. Other wireless gaging techniques that have been utilized include the use of inductive data transmission across an air gap and infrared light.

Figure 5.7 Touch-trigger probe with hardwired interface connection (courtesy of Martin Marietta Energy Systems, Inc.).

Condition Sensors

While position sensors are able to describe the dimensional condition of an object they are not generally useful for describing the other types of characteristics that are associated with the condition of a part or process. Condition sensors are used to measure information about the status of an operation (the dimensional data could be considered a subset of this larger group of data). Typical examples of this status information include the vibration level or frequency of a system component, the flow rate of a fluid, ambient atmospheric temperature and humidity, and the current flowing through a motor's armature. A brief description follows of a variety of sensors that can be utilized to define the condition or status of a process.

Photoelectric Sensors.   These transducers are among the most widely used sensors in industry [9]. They are used to detect the presence, absence and/or character of objects through the interruption or completion of a light beam. These solid-state devices are very reliable, with no moving parts or mechanical contacts. The photoelectric device that is most commonly recognized by most people is the original "electric eye." This type of unit has a separate transmitter and receiver and permits relatively long sensing distances. Other types of systems house the transmitter, receiver and power supply in a single package as small as 3 in. long and 3/4 in. in diameter and depend on a reflector to return to the emitted light beam back to the receiver. Also, fiber-optic photoelectric sensors may be used for the detection of parts as small as 0.040 in. In addition, most modern photoelectric sensors utilize an infrared light emitting diode (which reduces the problem of ambient light interference) that is pulsed on and off rapidly. The pulse modulation provides a high peak output from the LED without a damaging heat buildup.

Reflected Light Sensors.   These systems are utilized to measure surface form and surface texture. A machine was shown earlier, in Figure 5.1, in which a laser beam was focused to a beam diameter of 0.0004 in. and reflected off of a workpiece surface. This focused beam provides a measure of the average characteristics of the part surface within the given spot size. In addition, through the appropriate filtering of the system output signal an estimation can be made of form or surface texture. An alternative approach is to reflect the light off of the workpiece surface at an angle onto a photodetector array. The change in position of the reflected beam on the detector surface, as the part is scanned, can be related to the condition of the part surface [8]. Another technique is to examine the intensity of the light that is scattered back from the workpiece surface as a function of the angle of reflection [11,12]. In addition, interferograms can be utilized to characterize the quality of a part [13] or the accuracy of a segment of an operation such as a tool setting process [14].

Ultrasonic Sensors.   While these sensors were previously mentioned for performing dimensional measurements, they also are able to provide defect sorting by sensing the signature of a reflected sound pattern [15]. For materials such as metal, ceramic, glass, plastic, wood and composites it is possible to differentiate between the sound pattern reflected off a good versus a bad part. In addition, tramp material that has been introduced inadvertently into an automatic part handling system can be detected and shunted aside. Also, different types of products such as phillips head versus slot head screws can be correctly sorted. The limitation to this sorting technique is that it cannot be used with materials that absorb sound such as sponge rubber or other products that have similar acoustic characteristics.

Fiber Optic Sensors. Fiber optics are usually thought of as a means of conducting light to remote locations or as a transmission media for communications. Characteristics of these fibers which are undesirable in a communications environment, such as the microbending effects which create problems in cabling, can lead to sensor possibilities for measuring forces [16]. Other products include fiber optic gyroscopes and hydrophones as well as systems for flight controls and engine monitoring in aircraft and security devices.

Tool Condition Sensing. Tool condition can be sensed by a variety of means that vary from examining the dimensional characteristics of a portion of the cutting tool to measuring a secondary parameter such as force, vibration or surface finish. Tool breakage is typically detected through the variation in a vibration or force signal [17] and is relatively easy to recognize because of the drastic change in cutting conditions that occurs. Tool wear or tool edge fracture is more difficult to detect because the change in signal levels is smaller. Traditionally, the wear on the flank of a tool has been examined optically as means of inferring the radial wear on the tool edge. More recently, the size and shape of the cutting tool has been measured at different points in the manufacturing cycle as a means of estimating tool wear. In addition, for some applications the gradual change in workpiece dimensions over a series of parts may be used to predict tool wear. Unfortunately, while different techniques exist for monitoring tool wear during the metal cutting process, the methods that are the most sensitive measures of tool wear depend on information that is collected during interruptions in the machining process. Also, another factor that compounds the detection problem is that the amount of tool wear that must occur before a tool is considered worn varies with different applications. In general, the tighter the tolerance, the less tool wear that can be tolerated, although this number can be altered with tool wear compensation systems.

Temperature Sensors. Thermistors and thermocouples are temperature sensing devices that are readily available for sensing temperatures to a small fraction of a degree over the temperature range encountered in most manufacturing operations (neglecting furnace or other high temperature applications). Special probe designs are offered for use in air or liquids and on component surfaces. In addition, the temperature sensors are available with "intelligent" transmitters (based on a local microprocessor included with the sensor) that offer linearization and remote scaling. Advantages of these systems include low cost, rapid response and a voltage output that is readily interfaced to a temperature control system. (For some high accuracy diamond machining applications, coolant temperature control systems have been constructed that provide accuracy levels of $\pm$ 0.001 to $\pm$ 0.01 degrees F.) A disadvantage that is associated with this type of sensor is the need for wiring to an external readout. This restriction means that use with moving objects is difficult. Infrared thermometers are portable, noncontacting devices with good sensitivity and measurement spans that can exceed 1,000 degrees F. In addition, they can be used with diagnostic computers in a thermal imaging system to detect malfunctions in components, such as printed circuit boards, that develop hot spots and a characteristic thermal pattern that permits detection of faulty units. A disadvantage of these temperature measurement systems is that they can be relatively expensive. Also, for some machining applications, these transducers may be difficult to focus in addition to being influenced by stray chips and coolant.

Vibration Sensors.   Vibration can be due to a variety of causes from bearing failure to spindle imbalance. Irrespective of the source, it is desirable to sense the vibration level so that the error conditions can be recognized and corrected. Accelerometers and piezoelectric sensors are available that can be mounted at various locations on a machine to sense vibratory motion in any or all directions with frequencies as high as several megahertz. One convenient feature of this type of sensor is that the high frequency vibration signals are usually transmitted easily throughout a machine's structure so that the sensor does not have to be located at a specific position to obtain useful information. One feature that must be considered in the use of these sensors is their sensitivity to other disturbances. A change in the temperature of the sensor may provide an erroneous signal to a monitoring system. Another difficulty can be obtaining a measurement arrangement that is sensitive to the signals of interest but is insensitive to background noise signals that occur in the same frequency range. Whenever possible, the necessity of extracting the meaningful information from a "noisy" signal is an additional complication that should be avoided.

Load Sensors.   The load parameter of interest may be a force, torque or power and it may be sensed under dynamic or static conditions. The process characteristics that can be evaluated with a load sensor include depth of cut collision (wreck), took condition, chatter and chip condition. Measuring the current supplied to electric drive motors provides an indication of motor torque requirements during different operations and may be accomplished through the use of an auxilliary serial port that is associated with the motor drive system. This motor current signal can also be used to provide information on system loads. With hydraulic drives it is possible to measure pressure in order to obtain data on the load conditions. Load cells and strain gages also may be used although they are sensitive to changes in the thermal environment. In addition, load cells generally have relatively short time constants which means that they may be difficult to use for those conditions in which a parameter must be monitored over a lengthy period of time.

Pressure Transducers.   The most common type of transducer used to measure pressure is the traditional analog Bourdon sensor. These gages are found throughout most factories and are usually only equipped with a dial indicator readout for visual monitoring although units are available that operate with AC or DC systems. A simpler go/no go type of pressure gage is available for use in those applications where it is only necessary to be sure that a pressure level is above or below a certain value. Other types of pressure sensors may be used to provide a continuous output signal over a specific range. These units include variable reluctance transducers, piezo-resistive integrated circuits and fused quartz transducers. The variable reluctance transducers utilize a diaphram that is connected to electrical coils to translate diaphram motions caused by pressure differentials into an electrical output signal. These devices are used primarily in the lower pressure ranges. The solid-state transducers offer wider pressure ranges as well as insensitivity to wide temperature extremes. Measurement systems with digital output signals are available that can be readily interfaced with the electrical hardware used in digital monitoring and/or control systems. There are readily available devices that can measure pressures that vary from vacuum to thousands of pounds per square inch. These electronic pressure transducers can be interfaced with "intelligent" transmitters that provide improved calibration over the operating range, as well as offering the option for inferring fluid flow and level from the pressure measurements.

Electrical Characteristics.   Voltage, current, impedance, and so on, are all electrical characteristics that are indicative of the status of a system and may be measured readily. The most common devices for measuring electrical characteristics utilize a voltage measurement that is translated into the desired feature (such as current or resistance) through a calibration factor. Other sensor systems are based on the measurement of the electric field that is created as an electric current flows through a conductor (such as a timing light for an internal combustion engine). Digital readouts and outputs are widely available through the use of analog to digital conversion circuitry, but the electrical sensors used with these systems are inherently analog devices.

Velocity Sensors.   Velocity transducers are typically rotary devices such as analog tachometers that are attached to the shaft of a drive motor. However, linear velocity transducers are also available using units that are analogous to a linear motor. In addition, digital systems also can be used to convert pulses per unit time into a rate signal. This permits the output signal from a device such as an encoder or a laser interferometer to be utilized to provide both velocity and displacement information. Flow rate sensors with analog or digital outputs also are available that measure the speed of fluids.

Sensor Interfaces

Sensing the status of significant operating parameters is the beginning step in the endeavor of being able to control and improve the quality of manufactured products. Earlier sections of this chapter have discussed the transducer elements that are available for measuring the various characteristics that define the status of a manufacturing operation. In addition, the use and analysis of the process information that is provided by these sensors has been discussed in previous chapters. However, before the transducer signal can be analyzed it must be transferred from the output of the measurement device to the equipment that is used for the display and/or manipulation of this information.

The data analysis system may be as simple as an analog or digital readout or as complex as a mainframe computer. The data transfer interface may be as simple as a set of wires connecting the output of one device to the input of another, or it may be as complex as a radio frequency transmitting system that sends analog data to an analog to digital (A/D) converter that is connected to a computer input port. In any event, the function of the sensor interface (data collection and data conditioning hardware/software) is to transfer the intelligence contained in the electrical output of the sensing system to the appropriate point in the analysis system so that the data manipulation can be accomplished. In order to achieve this objective, the measurement information must be in a form that is readily accepted by the electrical input circuitry of the data analysis system. There is nothing to gain from sending a 120 VAC signal to a 10VDC input circuit without intermediate signal conditioning. In addition, the mechanism must exist to transfer the data from the input circuitry to the point at which it can be utilized. If software is used to assist this transformation, then an application-specific protocol must be followed to maintain the integrity of the sensor output. The following sections are a brief listing of a variety of interfaces that might be utilized with process monitoring systems.

Analog Interfaces.   These interfaces are utilized to transform an analog signal, such as an electric voltage or current, into a form that is compatible with an analog readout or with a digital processing device (analog processors are rarely used in current manufacturing operations). If the analog signal is to be used with an analog system, then scaling or filtering may be utilized to condition the input signal. When the system requirements necessitate that analog voltage information be translated into a digital word, then an electronic device called an A/D converter is utilized. There are four common kinds of A/D converters available: the counter type, dual slope integrating units, successive approximation devices and parallel processing units. The counter type converter is the slowest as well as the least expensive, while the parallel conversion device is the fastest and the most expensive. The successive approximation type is the most widely used. When

significant in terms of the system response. The multiplexing technique also pro-
analog current information is utilized it is converted into a voltage before being
digitized.

Digital Interface.

Digitized information is transferred from one digital
device to another through the use of an input port (integrated circuit) that is controlled by hardware circuitry or a computer's software commands. In some instances, the digital information is sent to the input circuitry and is read periodically
by the receiving system through a "polling" arrangement. This means that the
system periodically examines the input circuitry to read the incoming data without
regard to when the information arrives. In other situations, an "interrupt" is generated by the transmitting system to tell the receiving system that the data is available and should be acted upon unless a higher priority activity is in progress.
This type of status communication between systems is termed "handshaking" since
it insures that no data is lost in the transmission process. Another type of handshaking can occur without the use of interrupt circuitry. In this case, the systems
simply trade status signals that define the condition of the data transmission but
do not cause interruptions in the normal schedule of events for either system.
Following the initiation of an input command, the digital information that is present
at the input terminals is loaded into the computer's memory and is available for
analysis.

It may also be necessary to transfer digital information to an analog system.
This is done when a computer is used to close a particular loop in an otherwise
analog servo system. In this case, a digital to analog (D/A) converter is used
to perform the necessary transformation. Essentially, each bit of the digital word
is given an analog weight and a summing circuit is used to obtain an accumulated
value that is the analog equivalent to the digital value. This analog output may
appear to be a continuous signal due to the speed with which the computer is
able to update the output information. However, it really is a series of discrete
step changes that occur each time a new digital word is sent to the D/A converter
connected to the system output register.

Signal Multiplexers.

These circuits are used to time share system
resources that are only required intermittently. These devices can be used with
digital or analog signals as well as serial data transmission systems. For example, in a combined analog and digital system, the use of a multiplexer avoids
the necessity of having to dedicate a single A/D converter to each analog signal
input. In addition, it also permits multiple analog inputs to be channelled through
a single digital input port under the control of the computer program. The software dictates which analog signal will be processed through the A/D converter
at a given point in time. Since a computer will only sample the status of one analog signal at a time, unless parallel processing is involved, it is not necessary
to dedicate a separate A/D unit to each input signal. Although there is a finite
time delay due to the time spent switching the multiplexer inputs this is rarely

vides better utilization of the input ports since a large number of signals frequently can be processed through a single input port, while still permitting different sampling rates to be utilized for different signals.

Sample and Hold.   This interface element samples the level of an analog signal and holds it at a constant level until the analog to digital conversion process is complete. Otherwise, an A/D converter could be trying to convert a constantly changing signal which would result in data integrity problems. While it is recognized that the level of the external analog signal may change immediately after it is sampled, at least the information that is collected is valid for a particular point in time.

Parallel Interface.   This type of circuitry is used so that all of the bits of a digital system (computer, printer, etc.) input or output word can be read into or out of the system at the same time. For an output port, the output data is transmitted from an internal memory register to a buffer register in the output port where it remains until a new output word is sent to replace the old word. With an input port the contents of the input word are transmitted from the input wires to an internal memory register within the system upon the execution of the appropriate input command. While the values on the input signal lines may change continuously, the value that will be obtained with an input command corresponds to the status of the input lines at the moment that the information is transferred from the input lines to the internal register. The advantages of this type of data transmission arrangement are speed and simplicity of programming to accomplish the data input. A disadvantage can be that a separate wire is needed for each bit so that the cables can become quite bulky when a lot of inputs are used, and transmissions over long distances are a problem due to signal degradation.

Serial Interface.   This circuit transmits one character at a time in a bit stream so that only two signal carrying elements are required for the data transmission. A start signal is sent initially to alert the receiving equipment that a character is on the way, then the remaining bits are transmitted in a specific time sequence until the entire character is defined. At the completion of the character transmission, a stop signal is sent to terminate the process. This simplifies the wiring and eliminates the problems of cross talk between different wires. The encoding/decoding hardware and software at each end of the serial data link are more complex than what would be required for a parallel interface where the timing of the data bits is not critical, and the data transmission rate is slower. However, most computer network communications are performed using serial interfaces that transmit the information over wires or fiber optics because the parallel transmission of the information would be impractical, although, communications between computers and their peripheral equipment (such as a printer) may be through a parallel interface.

Wireless Interface. Wireless data transmission techniques include infrared light, telemetry and inductive coupling. The basic idea in each case is to avoid wiring that would become entangled in a machine's moving members. The inductive coupling technique is the simplest but it is only useful for small airgaps. The telemetry and infrared data transmission approaches offer a much greater distance capability but at a greater complexity and cost, and these techniques may require batteries for operation.

AC Input. At times it is necessary to monitor the presence of an AC signal at some point in a system. One technique is to use an electro-mechanical or solid-state relay to control a set of contacts that can be monitored as a digital input. Another approach is to rectify the AC waveform and use it to control a light emitting diode (LED) that is monitored by a light sensitive device. An advantage to the LED approach is the electrical isolation that is inherent in the circuit design.

Input Scaling. In some cases, it may be a requirement to attentuate a signal to avoid saturating the input of the monitoring system circuitry or it may be necessary to amplify a signal to increase the useful operating range. In either case, operational amplifiers provide a viable approach for scaling an input signal to the required level. However, it is desirable to consider the signal-to-noise ratio when performing signal scaling operations. If excessive noise exists, then it should be reduced whenever possible. A difficulty that can arise is when the noise and measurement signal have similar frequency ranges. Then, filtering out the noise also destroys a portion of the signal information.

Signal Conditioning. The presence of noise on a data signal degrades the resolution and accuracy of the sensor information. One approach to alleviating this problem is to utilize a low pass filter to attenuate the higher frequency noise signals from the lower frequency data. If filtering techniques cannot be used successfully, then it may be necessary to attempt to supress the noise at its source or obtain the signal in a different format such as an amplitude modulated or frequency modulated form. Differential amplifiers may be used to eliminate ground-potential differences in various parts of a circuit and optical isolators may be used to isolate circuits from each other.

Pulse Inputs. Some sensors provide up/down pulses to indicate an increase or decrease in a measured characteristic while other devices just provide a single pulse to indicate that a specific event has occurred. If these pulses are being sent to a counter, then it is necessary to be sure that the counter can operate fast enough to accept the pulses without losing information, and that it has sufficient capacity to sort the necessary data without being saturated or overflowing. When these pulses are sent to a computer system, then it is necessary to be sure that the computer will recognize the pulse when it arrives. For pulses from devices such as push button switches, the pulse rate is slow enough and the pulse width is long enough so that the monitoring system can be assured of recognizing the information when a polling arrangement is used to periodically monitor the data lines. For shorter duration events, it may be necessary to utilize an interrupt circuit to be certain that a particular occurrence is recognized. For high pulse rate circuits, a counter is needed to accumulate the pulses so that the data can be periodically transferred to the computer system. As mentioned earlier, when counters are used, care must be taken to insure that this input buffer has sufficient capacity to accumulate all of the pulses that occur between the computer data-input cycles to avoid the loss of information.

A quad B.   This is a form of pulse output common to encoders, linear scales, laser interferometers and other position measuring devices in which the transmission of displacement data occurs through two signals that are 90 degrees out of phase with each other (quadrature). Both amplitude and direction information are contained within these signals although additional hardware is required to extract the data. The circuitry to decode this information is readily available, although it is slightly more complicated than what is required for up/down pulses.

Resolver/Synchro Interface.   This circuit provides the necessary excitation signals for a position measuring device such as a resolver or a synchro and accepts the analog measurement output signal from the position transducer. Then, it processes this signal and provides an output in the form of a digital word that is directly compatible with the digital input port on a computer. In the past, the discrete element circuitry required to accomplish this task was somewhat complicated; however, these signal processors are now available as a single integrated circuit. Some models also output an analog signal that is proportional to velocity.

Software Interface.   The items discussed previously have described the hardware entities necessary to interface the output of various sensors to the input of a data analysis system such as a digital computer. However, before the information can be manipulated in a computer system it must be entered into the computer memory. The software interface is the computer code that causes the data at the input port to be transferred to a location in the computer's memory. This may require only a simple input instruction for a digital input or a series of commands for a serial input. In addition, other processing instructions may be required for device "handshaking" and to translate the input information into a form that is needed for the internal computations.

Sampling Theorem

In addition to the various considerations involved in choosing an appropriate hardware interface, it also is necessary to consider another factor when dealing with the timing requirements for the software interface. This factor, called the *sampling theorem*, is involved because the signal monitoring systems that utilize digital hardware are based on the process of sampling continuous signals and approximating them with a series of discrete data points. This subject deals with the requirements for obtaining an accurate representation of a particular signal when using sampled data. The theorem states that it is necessary to sample the signal of interest at a rate that is twice as fast as its highest frequency component. This does not mean twice the rate of the highest frequency component of interest because higher frequency components will appear to be shifted into a lower frequency range by the sampling process. If higher frequency components exist, such as high frequency noise, then these signals must be attenuated through the use of filtering techniques before the sampling process is implemented. For example, if the analog signal of interest varies at a frequency of 100 hertz or less but there is a noise component present that varies at 500 hertz then it is necessary that the sampling system monitor this signal at a rate of 1,000 hertz or higher unless the noise component is eliminated. If the noise component is removed, then a sampling rate of 200 hertz is sufficient to extract the useful information from the analog signal.

Failure to maintain a sufficiently high sampling rate will result in a phenomenon called *aliasing* which means that erroneous data is being collected. This effect was demonstrated vividly in the old western movies in which a stage coach was moving rapidly across the screen. Because the shutter of the motion picture camera was not operating fast enough (the sampled data rate was too slow), the wheels appeared to be moving backwards at times when the stage coach was obviously moving forward. This happened because each frame of the picture caught the wheels at a point that was slightly less than a full rotation. The end result was that when the movie was shown on the screen the wheels would appear to be rotating backwards at certain times.

A similar problem can occur in sampling data with a digital data-collection system. If the sampling rate is too small, then at times a high frequency component will appear to be occurring as a low frequency component which produces an incorrect interpretation within the lower frequency range. An extreme example of the confusion that can result from this process is easily demonstrated with a sinusoidal waveform that has a frequency of 250 hertz and varies about a zero level. If the sampling system also has a sampling rate of 250 hertz, and happens to collect the sample at the time the waveform crosses the zero axis, then the sample signal will appear to be equal to zero at all times. If the sample is synchronized with another point on the sinusoidal wave, then the sampled data will give the appearance of having a DC characteristic instead of the true AC wave shape. If the amplitude of the basic sinusoidal waveform varies, then the sampled signal will detect the frequency of the amplitude variations but not the frequency of the basic signal (assuming the amplitude variations occur at a frequency of less than 250 hertz). In order to avoid the alaising problem, it is necessary that the highest frequency data present in the monitored signal determine the sampling rate or else a high frequency filter must be utilized to insure that there are no high frequency components that could bias the information.

Machine Vision Systems

A discussion of the overall integration that is necessary with vision systems provides a good example of the way sensors can be utilized in a manufacturing environment. The vision system is a topic that has received a lot of publicity as an element of various automation systems and as an example of the computer-aided intelligent systems that will exist in the factory of the future. It has also been claimed that computers that can see can give American manufacturers the edge they need in their struggle with foreign competitors [18]. While machine vision systems can perform inspection and control tasks to determine part presence, location, orientation, and so on, the equipment must be matched to the specific factory environment. Also, the measurement tolerances that need to be achieved must be considered in light of the particular application. In the case of very expensive products or critical safety implications, the vision system cannot be permitted to make a mistake in the inspection process. For other applications the cost of incorrectly evaluating a good or bad part may not be too serious. In addition, applications may range from high production rate presence/absence sensing systems used with relatively inexpensive products to sophisticated dimensional measurement systems that require several seconds to examine an expensive workpiece.

One advantage of vision systems can be that for specific applications, they can provide 100% inspection without significantly slowing down the part flow. In addition, this technique can eliminate the boring, repetitive operations that human workers are not well suited to perform. This is accomplished by inserting an object into the field of view of a camera and converting the resulting picture into a format that can be used with a computer (digitizing). This transformation of the information contained in the picture is accomplished by projecting the camera image onto a light-sensitive sensor. The light-sensitive sensor is constructed of an array of individual elements called pixels that may be individually sensed to determine the light intensity at each point in the picture. This analog data can be converted into a digital code using an A/D converter and stored in a computer for further analysis. In some instances, the pixel information is evaluated using a simple on-off format in which the light intensity is judged to be above or below a specific value. This technique gives a rather coarse description because of the limited resolution. Gray scale analysis methods in which the light intensity is divided into finer increments (such as 64 levels) offer much better resolution but require additional analysis for data interpretation. In this case, the analog output of the system sensor is effectively calibrated over a particular range to extend the useful resolution and accuracy of the measurement device.

In some applications it is necessary for a vision system to examine a picture and compare it with reference data records that define the correct orientation or status of an object. This type of task may be considered to be a pattern recognition activity. In other cases it is necessary to look for the existence of a flaw that should not be present. If the flaw can be characterized as causing an unusual surface finish or other visual clue then the vision system can be programmed to recognize the problem. Another potential task for a vision system is to determine the physical location or dimensions of an object. The resolution to which this can be accomplished depends upon the number of pixels in the sensor array and the effective field of view of the camera. If the camera optics project a picture with a field of view of 0.320 in. by 0.480 in. onto a sensor with 320 × 480 pixels, respectively, then the basic pixel resolution of the system will be 0.001 in. This means that the object dimensions or location cannot be resolved to a level finer than 0.001 in. unless sub-pixel resolution is achieved through the use of the gray scale technique mentioned above. However, even with the basic 320 × 480 pixels, this means that over 150,000 pieces of data exist for each picture. If 64 levels of gray scale were added to this system, then approximately 9.8 million data points would be generated for each picture. Therefore, the problem is not just to extend the resolution of the system to finer and finer levels, but also to develop techniques for handling the information that is being generated in a timely manner. Fortunately, while some application areas, such as the exploration of outer space or the collection of intelligence information from a satellite, may require all of the information contained in an entire picture to as fine a resolution as possible, this is rarely a necessity in the manufacturing environment. It frequently is possible to determine the needed information to control or characterize a manufacturing process by identifying a workpiece edge, the intersection of two planes, or the location of a feature. This means that only a subset of all of the potential data that might be extracted from the sensor needs to be analyzed by the vision system. As with most sensors systems, the problem is not just getting the sensor data, but knowing what to do with this information once it has been obtained.

One example of the use of a vision system in the automotive industry is a robot cell that is used to apply seam sealant to automobiles at the Austin Rover plant in Cowley, England [19]. This system utilizes stereo vision to direct robots that apply mastic seam sealant to automobile bodies. Cameras are employed to read a pattern which defines the type of automobile and to determine the physical orientation of the automobile within the frame of reference of the robots. Using the position information provided by the vision system, the robots are able to apply the mastic sealant on the car body within an accuracy of plus or minus 0.04 in. (1 mm).

References

1. Sensor Management for Manufacturing Automation, Micro Switch, a Honeywell Division (1986).

2. W. E. Barkman, Laser gage qualifies machines, *American Machinist, 124*: 5 (1980).

3. G. J. Siddall and R. R. Baldwin, Developments in laser interferometry for position sensing, *Precision Engineering, 4*: 4, (1980).

4. N. Ikawa, et al., Photoelectric displacement sensor with nanometer resolution, *Precision Engineering, 9*: 2 (1987).

5. J. V. Moskaitis and D. S. Bloomquist, A microprocessor-based technique for transducer linearization, *Precision Engineering, 5*: 1, 5-8 (1983).

6. R. L. Resnick, Capacitance-based surface metrology, *Sensors, 4*: 5 (1987).

7. P. D. Chapman, A capacitance based ultra-precision spindle error analyser, *Precision Engineering, 7*: 3 (1985).

8. W. Brenner, Displacement and Velocity Transducers (Digital or Analog) Utilizing Non-Contact Magnetostrictive Techniques, Proceedings of Machine Monitoring Sensors for Untended Manufacturing, Itascam, Illinois (1987).

9. R. McGeever, Using light as a detection method, *Sensors, 4*:2 (1987).

10. K. Mitsui, In-process sensors for surface roughness and their applications, *Precision Engineering 8*:4 (1987).

11. R. Brodmann, Roughness form and waviness measurement by means of light-scattering, *Precision Engineering 8*:4 (1987).

12. H. S. Corey, Surface Finish Form Reflected Laser Light, Oak Ridge Y-12 Plant, Oak Ridge, Tennessee, Y/DA-7579, (1978).

13. R. E. Sladkey, Handbook of Optical Testing Methods for Diamond-Turned Mirrors, Oak Ridge Y-12 Plant, Oak Ridge, Tennessee, Y/DA-8005, (1978).

14. W. H. Rasnick and R. C. Yoder, Diamond-Turning Tool Setting by Interferogram Analysis, Oak Ridge Y-12 Plant, Oak Ridge, Tennessee, Y-2198, (1980).

15. G. S. Kushner, Sound out defects with supersonic sorting, *Production Engineering, 34*:6 (1987).

16. R. P. Main, An interview with Alan Harmer on fiber optics, sensors, and emerging markets, *Optical Engineering Reports*, No. 42 (June 1987).

17. L. E. Stockline, New Developments in Tool Condition Monitoring, Proceedings of Machine Monitoring Sensors for Untended Manufacturing, Itasca, Illinois (1987).

18. S. N. Lapidus, Machine vision, eyes for the intelligent factory, *Production Engineering 34*:4 (1987).

19. C. Loughlin, Robots provide seal of success, *Automation, 35*:1 (1988).

CHAPTER 6

Process Monitoring

Introduction

The quality control technique of monitoring a process in real time to characterize its status has been discussed in previous chapters. The objective is to observe those process variables which have been proven to be the best indicators of the quality of the manufacturing operation. Then, undesirable shifts in the status of these key operating parameters can be sensed before the production quality suffers. This early warning of increased parameter variation or other excursions provides information that permits the appropriate corrective actions to be initiated in a timely manner. However, this is often a task that is easier to discuss in general terms than it is to implement in a production environment. One of the more difficult objectives to accomplish can be the isolation of the most appropriate process variables for use in the control strategy. To obtain a process model that is not prohibitively complex, it is necessary that only a few critical process parameters be selected for monitoring and/or control activities. Models that utilize a redundant multitude of process parameters that are relatively insensitive to the status of the operation, are neither more accurate nor cost effective. In addition, another consideration that should be evaluated in the design phase of the control/monitoring system is the use of in-process versus post/process measurements. Also, the trade-off involved in the use of direct versus indirect measurements of the process variables for estimation/control of the quality of the process must be fully understood and accepted.

The postprocess measurement approach is one in which products are certified "downstream," in a time and geographic sense, from the point of production. This procedure, implemented in a wide variety of forms, is the method that is most commonly used to characterize the quality of today's manufacturing activities. This technique is attractive in one respect as the workpiece is already completed. Therefore, it is not necessary to estimate the quality characteristics of the part based on other attributes. In addition, the part usually is relatively clean so that accurate measurements are readily achievable. The only requirement needed to assure an accurate assessment of the quality of the product is to measure the appropriate features to the necessary level of accuracy (as defined by the workpiece tolerances). Otherwise, inaccurate error characterization can result in good parts being rejected and bad parts being accepted, as covered earlier in the discussion of sampling plans. Also, attempting to alter the process operation based on inaccurate information or attempting to make corrections for feature errors that are below the level of the accuracy of the measurement system does more harm than good. If this incorrect information is used to implement operational adjustments the process variation will be increased above the normal background level. This is because a significant portion of the time these process adjustments are misguided and compound the problem rather than alleviate it.

Another drawback of utilizing postprocess measurements can be the time delay between the point of fabrication and the point of inspection. This time delay varies from one operation to another, but the assumption is that the measurement phase occurs at a stage of the operation in which it is already too late to take corrective action for salvaging the current part or parts. (Rework may still be an option in this case, but it is an undesirable mode of operation.) As might be expected, this technique is most appropriate for sampling a well established, in-control process that is fabricating relatively inexpensive parts. In addition, it is desirable that the inspection operations occur as soon after the manufacturing cycle as possible. The applicability of this technique to other situations depends on the cost and probability of producing one or more defective workpieces before the problem is discovered. The complications arising from the production of 1,000 defective marbles are much less significant than the production of one defective mounting bracket for an aircraft engine. In the first case, it is cost effective to just scrap a certain amount of the product. However, in the other case not only is the product expensive but there are significant product liability issues to consider.

In-process measurements are more timely in nature, although they are often more difficult to accomplish. This can be due to assorted physical constraints such as the presence of chips or coolant that obscure a location that needs to be inspected. Also, other complications can exist when trying to measure some process characteristics, such as surface finish or size, while an operation is in progress. Furthermore, the ability to react to the results of these measurements may be limited at a particular point in the manufacturing process. In an ideal situation, it might be desirable to measure tool wear as it occurs so real-time process corrections could be employed. However, this parameter is difficult to characterize while a machining cycle is in progress so that it is usually necessary to settle for a dimensional measurement that is obtained between machining operations. At this point, the workpiece is still loaded on the machine and corrective actions can be implemented relatively easily.

Whether or not corrective actions can be initiated immediately, it is desirable to be able to predict degradations in process quality before they become significant or at least to recognize a shift in the important process characteristics as soon as it occurs. The alternative is to risk making additional defective products while waiting for the first faulty part to be inspected at some later point in the fabrication cycle. Even though it may not be necessary or appropriate to take action on the current workpiece, the identification of a process shift frequently occurs in sufficient time to avoid making out of tolerance parts. In addition, even if the current part is unacceptable, this procedure avoids producing additional parts that are unfit for their original design intent.

The issue of whether to utilize direct or indirect measurements of specific process characteristics for estimating the quality of a workpiece, is somewhat complicated by the different environments that can exist within each manufacturing operation. Techniques that may be suitable in one instance may be totally inappropriate in other circumstances. The decision on what is the best approach to employ hinges on the ability of the indirect parameter measurement to predict the status of the product characteristic of interest. If both approaches are equally accurate predictors of process quality, then the deciding issue becomes other considerations such as the initial cost, maintenance factors, the impact on other portions of a system or even personal preference. Another facet to address is that, in the case of parameters such as melting point or tensile strength, it may not be possible to make a direct assessment of the part characteristic without damaging the workpiece. In this case, it may be desirable to use an indirect parameter to monitor the process on a routine basis but employ the destructive testing on a limited basis to provide a spot check of the indirect monitoring technique.

The determination of the most appropriate system parameters for use in monitoring a manufacturing operation can be a complicated process. Often, there is a large number of potentially important process variables with varying degrees of sensitivity to the quality of the operation. The selection of the best ones to utilize in the construction of a process model is a task that is initially guided by engineering judgment. However, the final determination must be based on results from empirical tests. Professional opinion is certainly valuable as a starting point but it must be supported by sound statistical testing. All too often things are done in a particular manner just because everyone knows "that is the way they have always been done" or because a dominant personality thinks that it is the best approach. The accuracy with which the monitored parameters are able to predict the outcome of the process throughout the potential operating range is an excellent indicator of the suitability of the judgment criteria. In addition, the exercise of demonstrating the validity of the choice of key process parameters may highlight the existence of other unrecognized parameters that are also of primary importance. In this case, additional testing is required to isolate these important variables and establish their correlation with the quality of the operation.

Figure 6.1 shows a simplified sketch of a numerically controlled turning machine which can be utilized as an example of what might be encountered in attempting to select the appropriate process parameters for use in a process model. This figure also shows the tool path employed to fabricate a sample part. The work spindle is used to rotate the workpiece while the cutting tool is moved on a plane that passes through the center of rotation of the spindle. As the tool edge intersects the area of space occupied by the workpiece, material is removed from the part surface. This action creates a figure of revolution about the axis of the spindle. To a first approximation, the shape of the cross-section of the part corresponds to the shape of the tool path that is used.

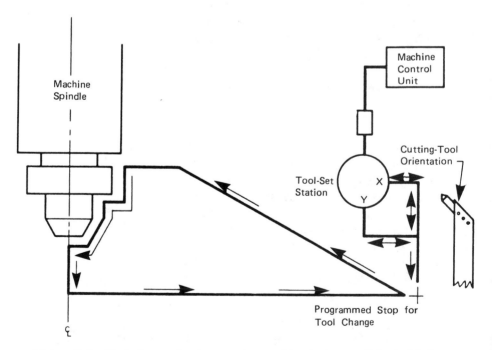

Figure 6.1 Typical operation for a turning machine (courtesy of Martin Marietta Energy Systems, Inc.).

The tool-set station shown in the figure is used to preset the cutting tool location prior to the machining pass. This is accomplished by moving the tool toward the center of the tool-set ball, using a single axis, until the tool-set electronics detect that contact has occurred between the ball and the cutting tool. This position datuming operation is necessary because the numerical control system is easily able to generate a programmed tool path, as represented by the arrows on Figure 6.1, but it has no way of knowing where to position the tool prior to the initiation of the machining pass. Errors in the starting position of the cutting tool produce errors in size features on the part, although the shape is generally correct. (The alternative to establishing the datum position for the cutting tool is to guess at the correct starting location and make a test cut. Then, the part can be measured and adjustments made for the subsequent machining operations. However, this is often an awkward approach, especially when multiple tools are used on a particular workpiece.) Figure 6.2 explains how the geometry of the tool-set ball, coupled with the tool-set procedure, allows the tool to be initially positioned by

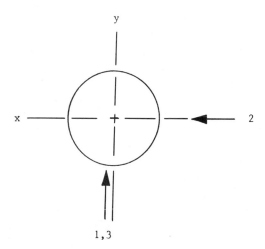

Figure 6.2 Description of tool-setting operations using a contact-ball tool-set station. Assume an initial tool position error of 0.1 in. in the X-direction, as indicated by the arrow at Point 1. This offset means that the tool will contact the 1 in. diameter ball at a location that is approximately 0.005 in. too close to the ball centerline in comparison to what would occur if the 0.1 in. error was nonexistent. This results in a .005 in. tool-set error in the y-direction, which means that when the tool is moved to Position 2, the tool is now misaligned with the ball center by 0.005 in. However, this only produces a tool-set error of about 13 μ in. in the x-direction. To complete the tool-set operation, the tool is returned to Position 3 to obtain the correct datum location for the y-direction.

"eyeball" and still obtain useful datum information for determining the tool-set location.

As described above, turning operations begin with the tool-set cycle. For the cross-slide axis, this establishes the location of the cutting tool in a radial direction from the center of rotation of the spindle. (This determines the dimensions of the part diameters since the closer the tool is to the spindle centerline at the beginning of the part program, the smaller the diameters will be machined.) Similarly, for the work slide, the important dimension is the location of the cutting tool edge with respect to the spindle face plate (which establishes part length). Next, the tool is moved to a predetermined location relative to the tool-set station, and the previously defined part program is executed to machine the workpiece. Assuming, for this simplified example, that the critical feature on this particular part is the large diameter, then a variety of factors exist that can influence the quality of the finished workpiece. These process variables include errors in the alignment between the machine axes and the work spindle, the properties of the raw material, tool wear and deflection, machine displacement errors, errors in the assumed location of the tool-set station and errors in the tool-setting operation.

As discussed earlier, a variety of approaches can be chosen to monitor the quality of this machining operation. It is intuitively recognized that some of the parameters such as spindle and axes alignment are relatively unimportant as predictors of the quality of the size control that is achieved on the larger part diameter. From the geometry of the operation it is recognized that the error in this variable is an insensitive measure of the part size control. However, this characteristic and others such as machine displacement accuracy do belong in the preventive maintenace category and therefore should be checked on a periodic basis. Other process parameters, such as the characteristics of the raw material, may be so important to the process quality that it is not possible to obtain a useful product unless these parameters are within rigidly controlled specifications. In this situation, it is desirable to be aware of the material status prior to beginning the machining operations. Unfortunately, it may not always be possible to make this determination ahead of time. If the problem is not recognized until near the end of the fabrication cycle, then the result can be poor quality control as well as poor productivity.

Other machining process parameters such as tool wear, system deflections and errors in the assumed location of the tool-set station all cause part size errors. The result of perturbations in these variables can be monitored by in-process inspection of the workpiece. In some cases, some of these parameters can be measured directly with on-machine gaging, while in other situations indirect measurements are obtained from the system sensors. The approach chosen is influenced by a number of considerations such as time, expense, ease of implementation and so on. Tool-setting errors and machine displacement errors are similar to small amounts of tool wear. In each case, the process characteristic is difficult to detect in real time. However, the effects of these error sources can be monitored through in-process inspection of the workpiece and machine offsets can be utilized to alleviate the problem.

The following sections of this chapter will discuss in more depth some of the different types of monitoring strategies that are available for a variety of manufacturing operations. In addition, examples will be presented that demonstrate the methods of implementation of these techniques.

Postprocess Monitoring

Postprocess monitoring implies a product evaluation sequence that is performed at a point in the manufacturing cycle that is relatively remote from the time at which the features of the finished workpiece were produced. At this location in the manufacturing cycle the parts have been removed from the machine and, in the case of small lot sizes, the machine may have been set up to produce a different workpiece. Examples of the types of product characteristics that might be checked include feature dimensions and locations, surface shapes, weight, hardness, resiliency, and so on.

One difficulty with this monitoring procedure is that rework of any features which are rejected in certification can be difficult to accomplish due to the problem of realigning the workpiece on the machine as well as the possible necessity of reconfiguring the machine for that particular operation. Also, an additional problem results from the evaluation of the quality of the product features at a relatively remote point in the manufacturing cycle. The difficulty can be that additional products are produced in the interim period between the time when a particular part is completed and when it is finally inspected. In the event of an excursion in the process, a number of deviant parts will be produced during the delay interval between the fabrication and inspection stages.

If the value of the manufacturing product is relatively small then it may be an acceptable to just scrap the defective units. However, this may create an undesirable atmosphere of apparent indifference to product quality. In any event, for expensive workpieces, it is necessary to have a high acceptance rate for the fabrication process. One way of accomplishing this objective is to employ a process which rarely results in a rejected feature. If this is not possible, then it is necessary to shorten the time interval between fabrication and inspection as much as possible. Then process excursions are detected more rapidly which effectively reduces the number of additional defective parts that are manufactured during the waiting period. An additional technique that is effective in this type of situation is the use of control charts. As mentioned previously in Chapter 4, control charts can enhance the early detection of process shifts so that the production of out of tolerance parts is avoided. However, a process for fabricating expensive parts which, under normal circumstances, utilizes most of the available tolerance band is like an accident waiting to happen. When this is coupled with a significant delay in the inspection cycle then the situation becomes quite precarious. Obviously, the smaller the normal process variation the better (within cost-effective limits) since this provides a buffer in the event of an unusual occurrence, but this is especially important in those situations in which the postprocess activities are delayed significantly.

One example of a relatively simple postprocess monitoring operation is a system to detect a faulty taped hole in a workpiece that is produced on an automated manufacturing line. This inspection operation usually occurs within a relatively short time after the machining cycle since it is necessary to evaluate the part status before further operations are attempted. The types of problems that could occur in this drilling and taping process are that the drill or tap could break and become imbedded in the material or that the hole or threads could be missing because of a damaged tool at a machining station. In the event that a defective part is detected, then it must be prohibited from continuing through the normal cycle. In addition, it is necessary to signal an operator that a malfunction has occurred so that the problem can be corrected. In a more sophisticated system, a tool change cycle can be utilized to correct the problem of a damaged tool without halting the process.

One method of detecting broken taps and drills is to use pin probes mounted on slides [1]. This process monitoring operation is performed in a testing station after the machining cycle is completed. A slide is activated that allows an air probe to attempt to enter the drilled and taped hole. If the hole is missing or an obstruction is present than the increased air pressure detects this problem. If the hole is clear, then the sensor is moved to the side of the hole where the absence of threads is also indicated by an elevated air pressure.

A more complex postprocess monitoring system would be involved in a gear manufacturing operation. Determining the quality of a helical gear is significantly more complicated than detecting the presence of a taped hole in a workpiece. Some of the part parameters that may need to be examined on a precision gear include eccentricity, pitch, profile, tooth spacing and helix angle. In addition, an attempt to rework this type of workpiece would be much more involved than just drilling or tapping a new hole. It is readily recognized that maintaining the normal process variations at a low level is extremely important since the process defects are only detected after it is too late to make an easy correction for a given workpiece.

In-Process Monitoring

As the interval (in terms of time and workpieces) between part fabrication and inspection is shortened the postprocess monitoring approach discussed above is more correctly defined as in-process monitoring. The optimal arrangement for this operation is to utilize direct, continuous, real-time measurements of those part attributes that are defined in the acceptance-tolerance criteria for the work-piece. The advantage to this approach is that the actual workpiece characteristics that will be used by the customer to judge the quality of the part are being moni-tored as they are generated. This process monitoring approach provides the most leeway in making system adjustments to compensate for undesirable product at-tributes. Unfortunately, this monitoring technique is often difficult to achieve due to processing constraints such as coolant, chips, inaccessibility of surfaces, and so on.

Instead of making these instantaneous process measurements an alternative method that can be utilized is to depend on measurements that are taken at nor-mally occurring interruptions in the production operation. For example, a work-piece feature, or other process characteristic, may be measured between machining operations to determine if the expected value has been achieved. While it still may be necessary to remove coolant or chips from the feature being examined in order to gain access to a surface that is to be measured, at least things are stopped in the sense that the coolant is off and rotating members are halted. Also, this technique has another advantage if the product is inspected while it is still mounted on the machine. Since the workpiece setup has not been disturbed, the part may be reworked or refined as necessary. In addition, this technique allows the cumulative effect of all the error sources to be monitored with one measure-ment operation. A disadvantage of this approach is that it requires the machine to be idle during the inspection process which may have a negative impact on total production rates (the total number of acceptable and unacceptable parts). However, it should produce a beneficial influence on the total production rate of acceptable parts.

The key factor to be concerned with in order to achieve the successful implemen-tation of this *process-intermittent monitoring* scheme is the repeatability of the process errors. This is because it is self-defeating to try to correct nonrepeatable errors. Also, the results achieved will be limited by the accuracy of the in-process inspection, the integrity of the data manipulation, and the resolution of the ad-justments that can be performed on a machine and the correction intervals used to adjust the machine operation. In some situations, viable alternatives to the process-intermittent approach are to inspect the workpiece on the machine im-mediately after the part fabrication is complete, or to utilize an indirect real-time measurement of part quality. Implementation expense is almost always an im-portant factor and in some circumstances the best approach may be a combina-tion of several of the above tactics. The most appropriate method typically varies widely for different operations based on the particulars of a given manufacturing operation and the product mixture.

An example of the successful use of real-time, in-process monitoring, in a manufacturing operation, is demonstrated by a system for the automatic tracking of a weld seam [2]. An important factor in achieving a high quality welding oper-ation is maintaining the proper standoff distance between the welding head and the workpiece seam even though the surface may not be perfectly flat. In order to monitor this process parameter, a sensor is utilized to measure the conductive properties of the metal that is being welded, which provides an indirect measurement of standoff distance. This sensor is mounted on the welding head just ahead of the area being welded so that it is not influenced by the weld puddle but it is sensitive to upcoming variations in the position of the material. Variations in the welding head standoff distance provide error signals that can be used to control the positioning servo system for the welding head. In effect, this is a closed-loop servo system which is analogous to the closed-loop positioning control systems used on machine tools. Machine tool position servos also provide continuous, real-time monitoring of a key process parameter called *position following error*. However, machine tool servos attempt to maintain a position following error value equal to zero while the above system attempts to maintain a specific, preselected standoff distance. (An additional complication for machine tools is that there are a number of other significant variables besides machine position that are not controlled, so that additional process monitoring is required.)

The use of opposed capacitance probes in a skiving operation for manufacturing fuses is another example of a continuous, real-time, in-process monitoring system for an untended manufacturing operation. In this instance, the capacitance probes are used as proximity sensors to continuously monitor the wall thickness of the fuse material [3]. A skiving tool is a type of flat form tool that can be used for metal removal on long parts. During the fabrication of some electrical fuses the skiving process is used to remove metal from the center section of a continuous metal strip that is later cut into individual fuse segments. The amperage rating of the fuse is determined by the thickness of the grooved section that is formed by the skiving operation. A typical dimension for the groove on one particular type of fuse is 0.125 in. wide and 0.013 in. thick with a thickness tolerance of plus or minus 0.0001 in. Since tool wear results in varying groove dimensions, the capacitance probe-based gaging system is used to monitor the thickness of the metal strip as it comes out of the skiving machine. The opposed head transducers are used to sense simultaneously the opposite sides of the metal strip which provides two signals that may be summed to generate a wall-thickness measurement that is independent of the vertical motion of the strip. This monitored parameter is used to adjust the tool position as a means of controlling the output quality of the operation. A gaging block holds the capacitance probes in position and controls the motion of the metal strip. Rollers are used to minimize the motion of the strip in the gaging direction and to insure that it remains within the measurement range of the displacement transducers. If the monitoring system becomes incapable of compensating for the tool wear then the skiving machine is halted and the system operator is notified of the problem. As in the previous example, a monitored parameter is used in a feedback control system to maintain the output quality of the manufacturing system. However, in this case, the monitored parameter also provides a direct measurement of a characteristic that is suitable for use in an acceptance specification.

The two previous examples demonstrate a monitoring strategy that accomplishes the goal of evaluating the part quality in real time without adversely affecting machine utilization. These applications required only the measurement of a single process variable to control the quality of the product. In more complex situations it may be necessary to measure a wider variety of parameters in order to obtain an accurate estimate of the quality of the process output. In addition, a combination of direct and indirect parameter measurements may be required in order to adequately characterize the process. This usually occurs in operations such as machining where it is desirable to measure the dimensional features of a workpiece. Unfortunately, it is not always easy to accurately gage the part surface in real time due to the presence of chips and cutting fluids. However, it is frequently a viable option to estimate a parameter of interest through an indirect measurement and at least be assured that no excursions have occurred during critical phases of the fabrication cycle. Two examples of machining process parameters that can be used for indirect measurements are the temperature of the machining cutting fluid and the vibration level generated during the machining operation.

The temperature of the cutting fluid used to flood the part during machining has a direct influence on dimensional quality. The temperature of this fluid is usually only marginally related to the shop air temperature. This fluid is heated as it passes through pumps and cooled by evaporation as it splashes over the machine. Measurements taken on precision finishing machines, in an air-conditioned shop, demonstrate that the cutting fluid temperature can easily fluctuate as much as eight degrees above the ambient air temperature variations over a 30 day period. While this temperature range might not appear to be excessive, experience has shown that a significant improvement in workpiece quality is achieved when the temperature difference between the shop air and machining coolant is reduced. The reason for this improvement in quality is that intermittently subjecting the machine members to the higher temperature collant sets up thermal gradients that produce stresses and distortions in the machine tool, while a cutting fluid that is near the ambient air temperature does not disturb the steady-state condition of the machine.

Figure 6.3 shows the inspection results obtained from the outer contour of two groups of eight-inch diameter hemishell workpieces that were fabricated on a precision lathe using different levels of coolant temperature control. The points plotted in the figure are the average size errors of the workpieces in each group. The inspection locations used to obtain the contour error data for the figure were a series of latitudes that were located at points between the pole and the equator of the part. (The error at the pole was zero in all cases because it was the reference datum for the measurements.) These parts were machined with all parameters held constant except that one group of 12 parts was fabricated with controlled temperature cutting fluid while the other group of 11 parts experienced the normal coolant temperature conditions. The range of temperature variation experienced

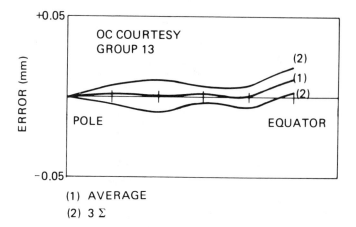

(1) AVERAGE
(2) 3 Σ

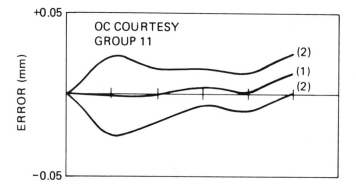

Figure 6.3 Contour inspection results for hemishell workpieces (courtesy of Martin Marietta Energy Systems, Inc.).

by each group of parts was approximately the same. However, the baseline coolant temperature for the control group varied about the nominal room temperature while the temperature of the other group varied about a point that was three degrees higher. The result was that the machine with the higher temperature cutting fluid was being periodically subjected to an intermittent thermal shock that was causing a degradation of its dimensional stability. From the inspection results, it is readily apparent that the lack of control of the coolant temperature had a negative effect on workpiece quality, and that this parameter is a prime candidate for being monitored and/or controlled.

Acoustic emissions and machine vibrations are two process parameters that also can be monitored in real time without disturbing the manufacturing operations. As mentioned in the previous chapter on deterministic manufacturing, the acoustic emission signal from a drilling operation has been used successfully to predict incipient drill bit breakage. Another source of acoustic emissions on a machine tool is the chip breaking process. Chip formation is an important process parameter for untended machining operations since long stringy ribbon chips tend to ball up and cause machine malfunctions. Insufficient chip control requires operator intervention to alleviate the situation. Acoustic emission signals have been used to detect chip breaking in aluminum under a variety of cutting conditions [4]. Unfortunately, this approach is unsuitable for use with materials such as stainless steel and uranium, especially when the small depths of cut are involved. The difficulty with these materials and conditions is that the acoustic emission signals do not vary significantly from the condition when the chips are being broken to when they are not breaking. Another drawback to utilizing this technique in this particular instance is that the normal machining operations produce noise artifacts with broadband vibration signatures that can be difficult to separate from the acoustic emission signals. Similar problems may be encountered in utilizing acoustic emission signals to detect tool breakage, although this technique has been used successfully to detect drill breakage.

The measurement of machine vibration is a technique that has been available for some time to detect problems in rotating machinery such as a faulty bearing or other drive train component. Equipment is readily available that stores a vibration pattern that is obtained as an indication of the machine's baseline condition when the machine is operating correctly. Then, the system compares this signature with subsequent measurements to detect trends that are indicative of component wear and the need for preventive maintenance. This is accomplished by using vibration transducers that convert mechanical motion into a proportional electrical signal that can be utilized to determine amplitude, frequency and phase angle. For most basic machine monitoring activities, these vibration characteristics will be sufficient to highlight problem areas.

This monitoring approach also may be applied to the problem of detecting tool edge fracture (the loss of a significant portion of the cutting edge in a localized area). Figure 6.4 shows the vibration signals that were measured while machining the order contour of three hemishell workpieces on a precision turning machine [5]. The vibration transducer (an accelerometer) was attached to the machine boring bar at a convenient point approximately six inches away from the cutting tool (although locations farther away from the tool seem to work about as well). The turning operation began at the pole of the part which is the area of minimum material removal per unit of time so the vibration level was also at a minimum level. Then the cutting tool was moved, in a circular arc, to the equator of the part. Because the orientation of the tool was fixed with respect to the machine axes, the point of contact between the tool and part changed constantly along the

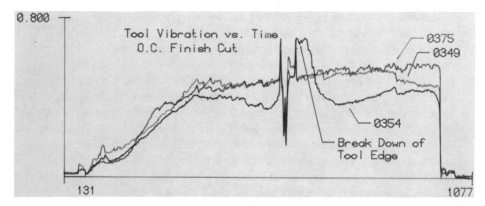

Figure 6.4 Vibration level sensed during a finish machining operation (courtesy of Martin Marietta Energy Systems, Inc.).

tool path. This meant that a new portion of the cutting edge was continually being used to remove the material. The excursion in the vibration levels near the middle of the signal trace is due to a special feature that is machined onto the outer contour of this particular workpiece. This feature causes the tool to plunge into and away from the part surface which results in the perturbation in the signal levels. It is easy to visually recognize that the trace for part 354 is significantly different from those for parts 349 and 375. Postmachining dimensional inspection of the workpieces revealed that excess material was left on the part surface of part 354 in the region near the midpoint of the tool path. In addition, examination of the cutting tool showed that the edge had fractured in the area that was used to machine the middle portion of the part. Ideally, this anomaly would be detected before the part was removed from the machine so that the tool could be changed and the part remachined. However, in this instance the vibration data was only being collected as historical information and not as a quality index for process control. The only on-machine monitoring that was being performed, for control purposes, prior to the removal of the part from the machine was the workpiece diameter. Unfortunately, this measurement was made at the part equator which happened to be a location that was not affected by the localized tool edge fracture. One problem that may be encountered with this technique is that detection of the loss of a relatively small portion of the tool is much more difficult to accomplish than recognizing the sudden loss of the entire cutting edge. Fortunately, the implications are also less severe. In the first situation the level of signal change may be fairly gradual as shown in Figure 6.4 while a more severe tool break is more often characterized by very abrupt changes in the waveforms. Finally, Figure 6.4 also highlights the necessity of comparing the current vibration pattern with a pattern that has been obtained under normal machining conditions. The presence of spikes in the measurement signal is not necessarily cause for alarm.

The previously mentioned process-intermittent inspection technique is in contrast to the continuous, realtime monitoring methods just discussed in that system-performance information is only available on a piecemeal basis. This approach employs process parameter measurements that describe the quality of the process at a point in the manufacturing operation that occurs prior to the final fabrication step. An example of this process monitoring approach is the manufacturing procedure used with the large diamond turning lathe [6] shown in Figure 6.5. This high quality turning machine is designed to fabricate a variety of large precision metal mirrors, using a single crystal diamond cutting tool. Diamond turning is

Figure 6.5   Large diamond turning lathe with sample test part (From Ref. 6).

similar to conventional turning operations in that a workpiece is mounted on a rotating spindle and the cutting tool is moved through the particular tool path that is required to generate the desired workpiece contour. However, significant differences exist between the tools used for conventional turning and those used for diamond machining. Diamond has the highest known wear resistance and the lowest coefficient of friction of any cutting tool material so that there is little edge wear and the chips slide across the tool face without causing a built-up edge [7]. In addition, in comparison to other cutting tool materials, diamond has the lowest thermal coefficient of expansion coupled with the highest thermal conductivity. This means that any heat that might be generated at the tool part interface is carried away from the cutting edge without causing tool expansion. When these tools are used on a high precision machine tool, the result is parts with mirror-like (optical quality) surfaces. One of the primary measures of goodness for these parts is the contour accuracy, or shape, of the mirror surface. In-process inspection and error compensation are used with this machine in order to obtain the optimum workpiece quality. The in-process inspection procedure used is to characterize the part contour using the on-machine, optical inspection set-up shown in Figure 6.6. This monitoring operation is performed at a normal pause in the metal removal cycle that occurs just prior to the final machining pass.

This use of the optical inspection method shown in Figure 6.6 provides a non-contact measurement of the part shape without risking workpiece damage. In addition, it does not depend on the accuracy of the machine slides to probe the part surface, as would be the case with more conventional on-machine gaging techniques. Another advantage of this in-process inspection approach is that the parts can be characterized while the work spindle is rotating (although it is necessary that the part surface be cleaned of cutting fluids, chips, etc.). This means that it is possible to evaluate the differences in part shape at different spindle speeds so that the effects of factors such as centrifugal force or workpiece imbalance can be studied. An obvious disadvantage to this inspection technique is that the equipment is rather bulky and it can be a time-consuming process to set up the equipment for the measurements (which has a negative effect on machine utilization). However, since the tolerance requirements for these workpieces are so stringent, the time spent in obtaining the process-feedback data is well worth the effort.

The device used to perform the in-process optical measurements is called an *interferometer*. This instrument measures length by using variations in the intensity of a light beam that are created by phase differences. The interferometer utilizes a beam divider that causes a beam of light to be split into two distinct beams that travel separate paths before being reflected back to the beam divider where they are recombined. One beam is sent to the part surface while the other is sent to a reference mirror. Differences in the shapes of the two surfaces result in phase differences in the reflected light beams which generates an interferogram (a pattern of light and dark lines). This interferogram can be evaluated visually or through the use of a computer program as a means of establishing the quality

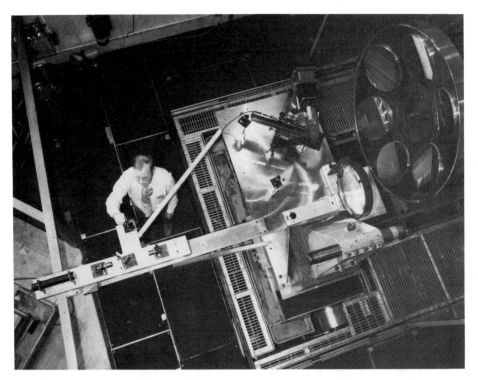

Figure 6.6   On-machine inspection of off-axis mirrors (from Ref. 6).

of the shape or surface texture of a workpiece. Figure 6.7 shows an interferogram of a mirror that was fabricated on the machine shown in Figure 6.5. In this case the fringe spacing is approximately 12.5 millionths of an inch. By tracing the fringe pattern (as shown by the dark line in Figure 6.7) and counting the overlapping fringes, it can be seen that the contour deviation of this off-axis paraboloid mirror is approximately 12 fringes or about 150 microinches. This *error map* can be used to adjust the final machining part program to correct for the repeatable errors. Obviously, this approach is only viable as long as the errors encountered in the semi-finish inspection stage are directly related to the errors that will occur in the final part (assuming no adjustments are made in the commanded tool path). In order to achieve this situation, extensive efforts are taken to maintain the machine's environment in a constant condition. Figure 6.8 shows an interferogram of the same mirror after it had been remachined with a programmed tool path that was modified based on the information gained from

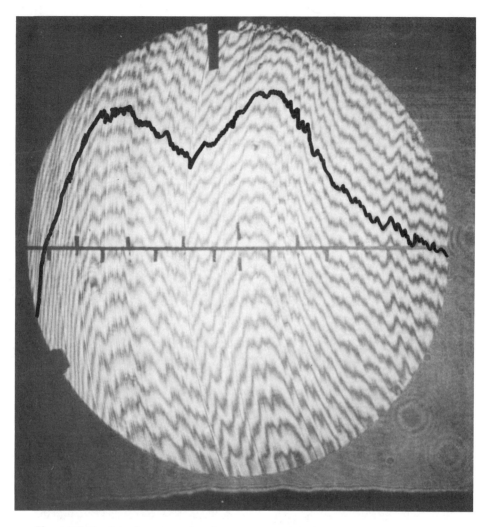

Figure 6.7 Interferogram showing part contour prior to compensation (from Ref. 6).

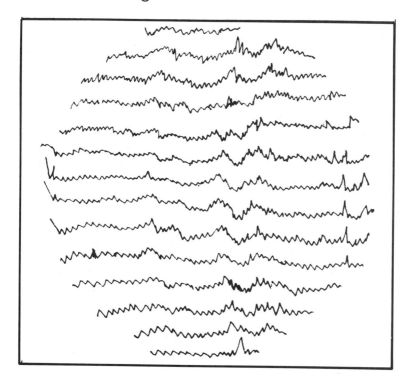

Figure 6.8   Interferogram showing contour improvement (from Ref. 6).

the earlier inspection stage. In this case the contour deviation was less than a fringe for an improvement in the workpiece contour of better than ten to one.

The above application may appear a little exotic to many people, but the same technique also can be valuable in other situations which require less demanding tolerances. One example of the application of this process-intermittent monitoring approach is a computer controlled contour grinding operation used to fabricate ceramic hemishell workpieces [8]. During conventional contour grinding operations, the machine's numerical control unit utilizes a part program to generate the necessary tool path. This part program is prepared using the assumption that the grinding wheel is a particular size and shape. While this may be generally true for single point tools, it is not necessarily an accurate assumption for all types of grinding wheels. This means that differences between the circular form of the actual grinding wheel cross-section and the size and shape that were assumed in the preparation of the grinding part program result in workpiece errors.

In this contour grinding example, a grinding spindle is attached to the tool slide of a two-axis lathe and the grinding wheel is moved along a tool path identical to one that would be employed if the workpiece was being fabricated with a single point tool of the same size. Figure 6.9 shows a typical grinding wheel cross section and the tool path that is employed to machine an inner and outer contour on a hemispherical workpiece. The grinding wheels are metal-bonded diamond wheels which means that, although they do wear when machining hard materials, they cannot be easily dressed to the desired size and shape. If errors exist in the grinding wheel form, this means that even though the grinding wheel is positioned properly to obtain the desired workpiece size at the equator of the part, there will still be shape errors on other portions of the workpiece contour. One method of dealing with this error source would be to prepare a different part program for each of the likely size values of the grinding wheels. Another technique is to inspect the grinding wheel cross-section to obtain a number that could be used in a conventional cutter compensation approach. However, neither of these potential solutions addresses the problem of wheel wear (which results in wheels that are the wrong shape as well as size) nor the likelihood of using the wrong tool path correction values with a particular wheel.

In this case, the solution chosen to avoid the above problems is to map the shape of the grinding wheel using an on-machine shape measurement station and to utilize this data to adjust the programmed tool path. The entire shape monitoring and tool path correction process is automatically controlled from the machining part program so that no operator intervention is required. In use, the machining part program automatically directs the characterization of the shape errors for

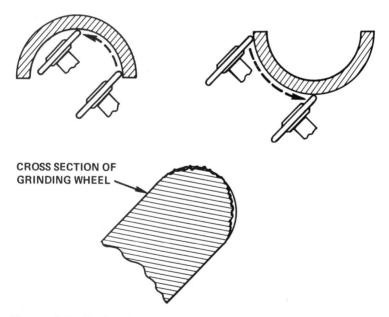

CROSS SECTION OF
GRINDING WHEEL

Figure 6.9 Tool path used for contour grinding (from Ref. 8).

the current grinding wheel and stores this data in an error table. Then, this information is used automatically during the machining pass to modify the theoretical tool path to compensate for the errors in the actual grinding wheel. Even in the worst case situation, where wheel shape errors exist that require tool path adjustments as large as 0.027 in. (perpendicular to the programmed path), the system is still capable of producing part contours with an accuracy of plus or minus 0.001 in.

Indirect-parameter monitoring may also be used at intermittent points in the manufacturing process to characterize a workpiece feature that is difficult to monitor in real time. Figure 6.10 shows a cross-section view of one type of machining fixture for finish turning the inner contour of a hemishell workpiece (the part previously has been finish machined on the outer contour). The goal of this final machining operation is to produce the desired contour on the inside of the part and to achieve a specific value for the wall thickness at the polar region. While it is possible to make a direct on-machine measurement of the size of the inner contour it is difficult to perform the same type of measurement for the workpiece

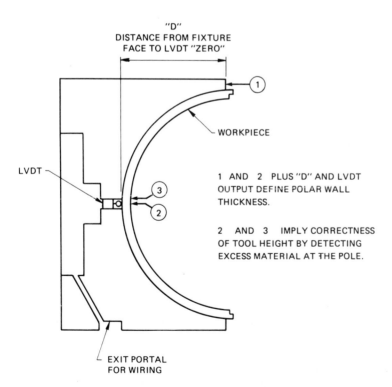

Figure 6.10 Cross-section of instrumented fixture for turning hemishell shaped workpieces.

wall thickness. This fixture depends on an external vacuum source to hold the part in place. As long as the outer contour of the part closely matches the inner contour of the fixture the workpiece will be held against the bottom of the fixture by atmospheric pressure, and a one-sided inspection procedure can be used to approximate the wall thickness feature. The problem occurs when there is a mismatch between the part and fixture contours (due to size or roundness errors) so that the part does not sit securely against the bottom of the fixture. When this happens a thin wall will be produced, without warning, in the polar region. This is because the pole of the fixture is the positioning datum for the tool in the direction of the work slide travel. An additional complication is that a part with this type of defect must be scrapped since it cannot be reworked due to a lack of material. The advantage presented by the fixture design shown in Figure 6.10 is that the position transducer (LVDT) imbedded in the polar region provides an indication of how well the part is seated within the fixture. Since the primary process adjustment required by the location of the part within the fixture is the offset of the machine axis, it is convenient to make this measurement while the machine is stopped prior to the final pass. Alternatively, if continuous measurements are desired it is possible to use a wireless transmitter to provide the position information in real time, although the use of this real-time data for process control may be inappropriate.

Another method of applying in-process monitoring is the situation in which the measured results at one stage are used to define the scope of the operation at a later stage. An example of this approach is a gaging system that is used with a different type of grinding operation to fabricate fuel injector parts for a diesel engine [3]. The tolerances on the inside diameter of the fuel injector pump barrel are plus or minus 4 millionths of an inch. The manufacturing system designed to accomplish this task automatically loads the workpiece, finish grinds it, and sorts the piece parts into containers according to size. In each cycle, the part barrel is gaged and a calculation is performed to determine the amount of material that needs to be removed in order to obtain the desired workpiece size. Based on this in-process feedback information, the part is automatically remachined and reinspected. This integrated system eliminates the need for a dedicated machine operator, reduces scrap, provides accurate production data and eliminates wasted time on piece parts that are not usable. However, this operation is different from the ones discussed previously in that the grinding-inspection cycle continues until the part-size tolerance is achieved or a useful part can no longer be obtained.

An alternative approach to the variety of monitoring techniques previously discussed is to perform the workpiece measurements after a first part is completed but before a second one is finished. In this situation, it is assumed that the process is stable and that only minimal changes are likely to occur from one part to the next. Therefore, once the appropriate operating parameters are obtained it is only necessary to fine tune the process. In addition, if an excursion does occur in one or more of the key process parameters, then adjustments can be made in time to salvage the following workpiece. This has the advantage of permitting the measurements to be made off of the machine but it does not necessarily optimize the quality of each part. Also, while the machining conditions/parameters may be maintained in a constant status from one part to the next, there is the possibility of variations in other factors such as the mechanical properties of the workpiece. If this occurs then the process parameter values that were appropriate for one workpiece may not be suitable for the next one. The end result can be to degrade the quality of the finished parts through the use of inappropriate process adjustments. However, in those situations where the process parameters are known to vary slowly or the impact of an unexpected process shift is not drastic, this can be the most suitable avenue to follow.

Process Models

The goal of the process monitoring procedure is to estimate, as accurately and rapidly as possible, the quality of the output of a particular process or group of processes. In order to accomplish this, certain process parameters or variables are measured as a means of evaluating the actual quality of the finished workpiece. Whenever possible the desired product attributes are measured directly. However, if this is impractical, then an indirect measurement of an appropriate feature is utilized. The factor which is critical to the success of this technique is to obtain an accurate system model for representing the performance of the system under the appropriate operating conditions. Then, it is necessary to monitor only those variables which are of major importance to the process quality. Secondary variables also may be monitored on the basis of curiosity or for a variety of other reasons. However, they should not be utilized in the estimation of process quality because this would add further "noise" to the measurement process.

The model used to transform the process measurements into an accurate assessment of the process output should be as simple as possible without losing the needed sensitivity. The addition of extra secondary parameters complicates the model, adds increased uncertainty to the results and is a extra expense with negative return. For example, including a realtime tool path accuracy parameter in a model of a machining system might be thought to be important. However, a good tool path does not necessarily result in a good part so this is not as useful a parameter as dimensional measurements of the workpiece quality. While it might be useful from a trouble shooting viewpoint to have this information available, it is redundant from a process monitoring standpoint. In addition, it adds unnecessary complexity and cost to the monitoring system. Given a particular model, the best way to evaluate its effectiveness is to conduct experimental tests to determine the model's validity for the range of operating conditions that are likely to be encountered. This provides experimental data about the sensitivity of the model and the physical system to the status of the various parameters.

At times, the physical phenomenon being modeled are relatively well understood and it is possible to write a descriptive equation that defines the operation of a particular manufacturing operation. Considering the numerically controlled machine tool/workpiece combination described in Figure 6.1, it is relatively easy to postulate a theorectical model that predicts part size in response to certain process parameter measurements. For example, one approach would be to state that the size of the large diameter on the finished part is a function of the physical position of the tool cutting edge at the beginning of the part program, the mechanical deflections that occur in the operation and tool wear. Unfortuntely, in this instance, it is easier to define the system variables than it is to measure them in real time. Since these individuals parameters cannot be measured readily it is futile to hope to make individual process adjustments for each of them. Fortunately, a more reasonable approach is to monitor the composite effect of the variations in these individual variables. Rather than attempting to monitor each individual parameter and theoretically calculate a cumulative error factor, the more sensible tactic is to make use of direct measurements of the diameter of interest. This can be accomplished through the insertion of a gaging cycle, that employs an on-machine probe, at specific points in the fabrication cycle. One viable technique involves the on-machine measurement of the size of the part before the finish machining pass in order to determine an appropriate offset for the finish pass. This alteration in the staring point of the cutting tool compensates for the error conditions that were in effect during the prior machining pass (which resulted in an erroneous workpiece diameter). This simplistic model works well as long as the system operating conditions remain relatively constant. In effect, what is being assumed is that for a limited operating range (two successive machining cycles) the system response can be predicted using only the measured diameter of the workpiece at a time just prior to performing the last machining operation. This particular model is used in conjunction with the process-intermittent diameter measurement to control the output quality of the machining operation. It is called an empirical model because it is based on observations of the process as opposed to the theoretical considerations. Other models, utilizing the dimensional information, also could be used to monitor the condition of the workpiece after the final machining cycle and to generate a flag if an out-of-tolerance condition existed.

A slightly more complex error model could be used in this situation to predict the quality of the final product prior to the last machining pass. In this case, the error estimation equation might take the form:

$$E_2 = C_1 + C_2 [DOC_2 - DOC_1]$$

where DOC_i is the depth of cut for the first and second machining passes and C_1 and C_2 are constants. The constant C_1 represents the combined uncertainty involved in the in-process measurement of the control dimension on the workpiece and the determination of the appropriate machine offset. The assumption is that an error could be made either in the measurement of the part size or in the calculation employed to determine the necessary machine offset for the final metal cutting operation. The term involving C_2 accounts for the difference in the depth of cut variable between the first and second machining passes (it is assumed that all other variables are unchanging). Then, E_2 is the predicted final error that would result at the control location on the workpiece (based on the condition of the part just prior to the final machining operation). This model is intended to function accurately over a linearized area of operation in which the depth of cut variation between the two cycles is small so that the machining forces are essentially constant. In addition, the baseline assumption is still that the other environmental factors that have the potential to affect the system operation are also relatively constant. In effect, this model predicts that if the depth of cut is essentially the same for both operations, then the only error in the finished part dimension will be due to the inaccuracies inherent in measuring the part and determining the appropriate offset. This is too simplistic a model for most real world machining operations, however, it demonstrates the basic concept in addition to highlighting the fact that the degree of accuracy of the predictive calculations is dependent upon the completeness of the system model.

The degree of complexity that is required to achieve the desired accuracy in an empirical model may be difficult to estimate before system tests are conducted. In this instance, it may be useful to begin with a more general model that can be simplified after some experimental testing is performed. One example of a more general model for use in the previous situation would be

$$E_2 = C_1 + C_2 [DOC_2 - DOC_1] + C_3 [DOC_2 - DOC_1]^2$$

where provisions are included for a higher order term, if needed. The next step in the process is to conduct a series of experimental trial runs to determine if the model is an accurate representation of the physical situation. Erroneous assumptions may have been made concerning the necessity for the use of certain process parameter measurements as well as the manner in which these measurements are combined to produce the system model.

One problem in designing an experiment to test the accuracy of a model is the selection of the points to be included in the trial runs. In the above example, it is likely that an experimenter can estimate the probable range of values for the term $[DOC_2 - DOC_1]$. The remaining decision is how to space the test points within this range. Since most relationships usually exhibit a degree of smoothness it is frequently appropriate to spread the test points evenly over the range of interest and conduct a preliminary evaluation. Then, if it appears that there are areas which deserve additional examination, the experiment can be redesigned to pay more attention to specific locations within the test space. Another possibility is that the variables selected for the initial model are inappropriate or insensitive to changes in system output in which case it is necessary to redesign the model and continue.

A reasonably sophisticated model of machine tool accuracy is demonstrated by work performed at the NBS on a slant-bed turning center [9]. (This work was mentioned briefly in Chapter 3.) In this case, the goal was to obtain a general mathematical model that could be used to calculate the vector error between the part and cutting tool based on measurements from a large number of reproducible error sources. The general model was designed so that the total error could be determined from the individual error components that corresponded to the errors contributed by the machine structural elements. Then, a predictive machine tool calibration technique was employed to decompose the individual errors into their geometric and thermally induced components. The final model was generated by utilizing mathematical matrix manipulations, in conjunction with the assumption of rigid body mechanics, to describe the relative positional relationships between the various structural elements of a machine tool.

In order to estimate the error in the relative positions of the workpiece and cutting tool at any location within the machine work zone it is necessary to predict all of the individual error components. Although these errors are geometric error components of the structural elements of the machine tool, their characteristics change as a result of such factors as thermal effects and loading conditions. Thus, each error term, e_i, can be considered to be a combination of factors related to the nominal position, x, and the temperature, T. This can be expressed in equation form as

$$e_i = a_0 + a_1 x + a_2 x^2 + \cdots + b_1 T + b_2 T^2 + \cdots$$

The data used to perform this analysis are obtained by making error measurements while monitoring the nominal axis positions and the temperatures of various machine components. Figure 6.11 shows a temperature profile for a turning center in which the machine temperatures rose 8–12 degrees C (15–20 degrees F) during a 500 minute run. In this instance, these temperature increases caused errors of about 2 μm per degree C (40 μin per degree F) in the direction radial to the spindle and about 20 μm per degree C (400 μin per degree F) in the axial direction. In order to find the thermally induced changes in the machine errors, the position-error measurements were taken over the entire machine while the temperature of the machine elements was monitored. The temperatures of the major heat sources, such as drive motors and bearings, as well as the temperatures of thermal barriers, such as the interfaces between cast iron structural elements was recorded.

The error components associated with this machine were classified into four groups with similar characteristics, measurement procedures and measurement instrumentation. The groups selected were linear displacement errors, angular errors, straightness-parallelism-orthogonality errors and spindle thermal drift. Each of these groups was characterized using a least-squares regression analysis that provided the individual error relationships that were needed to determine the error

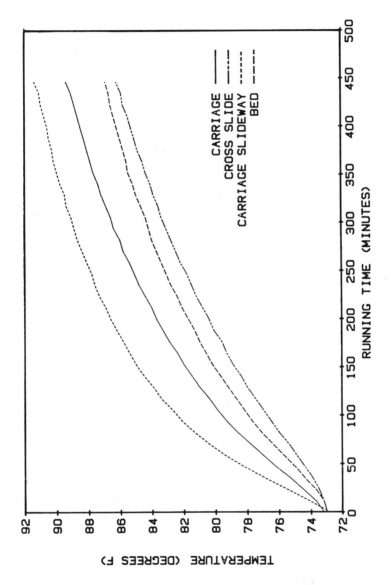

Figure 6.11   Temperature profile of machine elements under continuous running conditions (courtesy of the NBS).

vector in real time. Cutting tests conducted with and without the error compensation system showed that accuracy enhancement of up to 20 times is achievable with this technique. (Of course, the degree of improvement is related to the initial condition of the machine tool.)

An alternate approach that can be taken with the modeling process is to compare a system's existing performance with historical performance records. The vibration data shown earlier in Figure 6.4 is an example of this type situation. One or more baseline runs are made to establish an expected value for the system response at specific points in the manufacturing process. Then, this data is used to provide an on-line real time evaluation of the process performance. An advantage of this technique is that it is insensitive to the normal shifts in the information signal, as occurred in Figure 6.4, because only deviations from the baseline are considered important. Another technique for making the historical comparison is to utilize control charts, which were discussed in Chapter 4, to compare specific parameters at particular points in the manufacturing cycle. This method shares an advantage with the empirical model process in that an in-depth understanding of the process mechanics is not required. Instead, it is only necessary that the desired process variables be monitored during a period of stable operation to form the baseline characteristics of the system. A shortcoming of this approach is that an error condition that exists from the beginning of operations will not be detected; instead, it will be assumed to be a normal characteristic of the process.

Returning to the depth-of-cut models mentioned previously, it can be seen that the use of historical information also may be beneficial in improving the suitability of an empirical model. It may be desirable to have information about multiple points on the workpiece shown in Figure 6.1 because of the possibility of the potential for excessive tool wear. However, it is nonproductive to spend a lot of time with the on-machine inspection activities that can be involved in probing a large number of points. One approach to resolving this conflict is to measure a sample diameter on the part and obtain the tool wear informtion from another source that does not significantly impact the availability of the machine tool. As long as the tool wear was below a value that had been previously demonstrated to provide good products then the workpiece could be accurately evaluated using the limited on-machine inspection results. The tool wear information could be obtained through the use of an on-machine vision system or other sensor that quickly monitors the condition of the cutting tool at a normal interruption in the process. In addition, in some applications it may be possible to employ a vision system with sufficient range and accuracy to inspect the entire workpiece all at once. Therefore, it would be unnecessary to obtain the tool wear data to estimate the part quality. However, this information might still be needed to provide the necessary signal to initiate an automatic tool change operation.

References

1. Sensing broken tools on an automated line, *Sensors*, Nov. 1986, pp 20–21.
2. L. Garmaise, Sensor-based weld seam tracking, *Sensors*, Nov. 1986, pp 76–78.
3. P. Davis and S. Kilmartin, The use of noncontact capacitance sensors to facilitate untended manufacturing, Proceedings of Sensor Technology for Untended Manufacturing, Chicago, Illinois (1985).
4. D. A. Dornfield, The role of acoustic emission in manufacturing process monitoring, Proceedings of Sensor Technology for Untended Manufacturing, Chicago, Illinois (1985).
5. W. E. Barkman et al., Statistical process control for lathes, Proceedings of The Southern Manufacturing Technology Conference, Charlotte, North Carolina (1987).
6. H. L. Gerth et. al., Fabrication of off-axis parabolic mirrors, Proceedings of Society of Photo-Optical Instrumentation Engineering, San Diego, California pp. 93–102 (1978).
7. P. Donald Brehm, Diamond Turning and Flycutting for Precision, Society of Manufacturing Engineers Technical Paper MR77-965, Dearborn, Michigan (1977).
8. W. E. Barkman and T. O. Morris, Compensation for the Shape of a Contour Grinding Wheel Using Computer Numerical Control, Union Carbide Corporation, Oak Ridge Y-12 Plant, Oak Ridge, Tennessee, Y-2067 (1977).
9. M. A. Domnez et. al., A general methodology for machine tool accuracy enhancement by error compensation, *Precision Engineering, 8*:4 (1986).

CHAPTER 7

Computer-Controlled Manufacturing

Introduction

In the past, traditional manufacturing processes have involved a significant degree of manual operations. Initially, this was because the only available option was to do things by hand even though various types of manual mechanisms were available to make the manufacturing task easier to accomplish. More recently, automation has become another tool that can be utilized advantageously. This latest implement has become a driving force that is part of an evolutionary process in the manufacturing world. The result of this so called disturbance is a certain amount of continual change that occurs within existing manufacturing operations as well as an impact that is manifested in the changes that are incorporated into the design of new facilities.

At an earlier point in time, metal working was performed by a blacksmith laboring over a hot forge with a heavy hammer. Later, various machines were developed that relieved the human operator of the burden of providing the muscle required to shape a workpiece into a desired configuration. However, the skill of the worker in directing the actions of the machine was still the major factor in determining the quality of the finished workpiece. Eventually, machine control systems were developed that surpassed the human operator in terms of the ability to synchronize the various machine actions (such as the positioning of different axes or maintaining a constant parameter under varying conditions). Nevertheless, a single operator was still dedicated to a particular machine in order to perform the task of monitoring the machine's operations, loading and unloading parts, pulling chips, detecting malfunctions, and so on.

Today, systems employing digital computers are used extensively to accomplish work that was previously done in a manual fashion and the role of many human workers in the manufacturing industries has changed drastically. Instead of providing the "brute force" or "artistic skills" that were once required for many jobs, the worker's role has expanded in complexity and sophistication while moving away from the hands on mode of operation. Tasks that previously required varying degrees of special training such as the balancing of a servo system for a DC motor drive or moving the axes on a machine tool are being accomplished automatically through the use of computer systems. Now the craftsman is responsible for a much larger, more complicated manufacturing system and his skills are directed more at keeping the system operating properly than being concerned with the intimate details involved in fabricating an individual workpiece. (An exception to this characteristic is encountered when performing system diagnostics, but even these functions are becoming increasingly automatic.)

Examples of the equipment associated with the new generation of manufacturing systems include robot welders on automotive assembly lines and machine tools that are directed by computer-based numerical control units. The textile industry is another instance in which the application of automation techniques has been beneficial. In this case, manufacturing procedures have moved from the manual cutting of a single piece of material to the automatic cutting of multiple pieces of material. A common denominator associated with each of these situations is that increased levels of automation have resulted in an improvement in overall product quality, as well as a more desirable position in the marketplace.

Early automation efforts were largely based on the installation of assembly line types of systems that moved the product hardware automatically past a number of work stations. These stations were occupied by individual workers with a moderate degree of skill in the particular task needed at that specific step in the manufacturing process. Complex components could be manufactured economically although no single individual was necessarily trained to perform all of the tasks required to fabricate the complete workpiece. In addition, power tools were provided whenever possible (such as drills, cutters, screwdrivers, etc.) that tended to speed up the operations so that the physical stamina of the workers was no longer as important as it had been previously.

Although this mode of operation reduces the skill level required of the workers, it also introduces the potential problem of worker boredom due to the constant repetition of the task. While moving workers around among different stations tends to reduce the potential for boredom, it does not reduce the dependence on worker performance. The lowered skill level requirements do permit increased flexibility in worker selection, but the system is still paced by the performance of the workers and if someone has an off day then the entire process may suffer.

While this may be the best approach in some instances, especially in an application requiring relatively low skills coupled with relatively low labor costs, long term it is not the best way to utilize the innate talents of the employee (the ability to reason and be creative).

In some industries, this assembly line type of manufacturing approach has evolved to the point of replacing many manual operations with completely automated equipment. These automatic systems are designed to depend on digital computers and the resulting operations excell in consistently repeating the same job as well as offering a quality and production rate that are usually unmatched by manual techniques. However, when the computer-based system malfunctions (which will eventually happen) it also can have wide ranging effects unless provisions are included in the system design to adapt to this event. Also, the computer system generally requires a human operator to provide maintenance. In fact, the computer can be thought of as just another machine in the evolution of manufacturing processes and as such it is dependent upon a human operator. A creative operator can make this computing machine perform wonderous tasks while a less talented individual may create a monster.

The flexibility of the computer-based manufacturing systems varies among different applications. In some instances, where it has been possible to design the system to accommodate a variety of job types, the manufacturing versatility can be as great or greater than a human operator. In other situations, the complexity of the application requires significant manual intervention when a change over to another operation is required. At the end of this progression of capabilities is the situation, termed "lights-out" manufacturing, in which entire factories are able to conduct routine operations with only a smaller number of workers whose main tasks is to oversee the automated fabrication steps and provide needed maintenance. While these workers are involved in maintaining the status quo in the factory, they do not take an active part in the actual product manufacturing sequences. The key technology that permits this mode of operation to be successful is the use of interconnected, integrated, computer-based control systems.

Manufacturing computer control systems may be as simple as the single integrated circuit that resides on a printed circuit board and controls the operation of a DC motor or as complex as the hierarchical computer network that controls the scheduling, material handling, fabrication and inspection of a variety of workpieces. In keeping with the overall scope of this text to address the issues associated with process quality in discrete piece-part manufacturing, this chapter begins the discussion of computer control from a relatively simple viewpoint and moves to an increasingly sophisticated level. However, it should be stressed that technology, per se, should not carry too much significance. An important point to remember is that the solution to manufacturing problems must be an approach that offers cost-effective production of a quality product that is delivered on time and meets the required specifications, while meeting the long-term needs of the manufacturing organization. (While short-term requirements are important, it is sometimes too easy to become so involved with the solution to today's problems that future needs are overlooked.)

This chapter begins with a discussion of the origin of the machine control systems that have evolved into today's versatile manufacturing tools and continues with the explanation of the basic steps that this evolution has followed. The topics presented are set-point control, programmable logic controllers (PLCs), numerical control (NC), computer numerical control (CNC), direct numerical control (DNC), cell control systems, computer-aided manufacturing (CAM), computer integrated manufacturing (CIM) and knowledge-based systems (KBS).

Set-Point Control

Perhaps the simplest form of control device is an on-off switch such as is used to control an electric light. This type of control system is frequently called a bang-bang controller because it only has two possible conditions or states. It can also be thought of as a digital control for the same reason. A variable dimmer for an electric light is an example of an analog control system since continuously variable control settings can be obtained. However, both of these systems are open-loop control systems in that they do not alter their actions based on conditions such the amount of ambient light or changes in supply voltage. A closed-loop control system employs a feedback mechanism to sample a characteristic of a process and modify the control actions accordingly. An electric heater with a thermostat is one example of this type of system. As the area around the heater thermostat changes its temperature, the thermostat turns the heater on or off to maintain a relatively constant temperature in spite of the changing ambient conditions. In other words, a set point is maintained through the use of a bang-bang controller that varies the on-off times to achieve the desired temperature (within the capacity of the heater). An example of an analog feedback control system is the distributor on older automobiles. This mechanical mechanism varies the ignition timing continuously in response to changes in engine rpm and throttle position (carburetor vacuum) so that the best driving performance is obtained in spite of changes in operating conditions. The obvious advantage to the feedback control system is that it is able to maintain a higher level of control quality over a wider range of conditions than is achievable with an open loop control system.

As demonstrated above, the reason for using a control system in a manufacturing or other environment is to produce a system output that meets some prescribed conditions. Control systems are designed to examine the condition of a process and vary one or more parameters as needed to maintain the quality level of the output product. The process goal, or set point, is different for each application and may be a temperature, a force, a position, and so on. In addition, this set point may change with time in a manner that is defined by the requirements of the operation. For instance, the temperature of a furnace may need to be maintained at a level of 200 degrees centigrade for one hour followed by a rise to 250 degrees in 10 minutes and a decrease to 22 degrees in three hours. In this case, a computer may be utilized to store the desired time and temperature parameters and manipulate the furnace temperature controls to achieve the desired pattern of operation. The analog feedback signal is the temperature of the furnace, while the controlling action for the furnace heater may be an on-off or proportional analog type. The control signal is generated by comparing the desired time-temperature profile, mentioned above, with the actual temperature at a specific point in time. Deviations from the desired set-point profile cause the furnace heaters to increase or decrease their output levels in order to drive the actual temperature closer to the desired temperature.

Figure 7.1 shows a more exotic example of a set-point control system in which a computer is used to achieve extremely accurate control of the position of an air

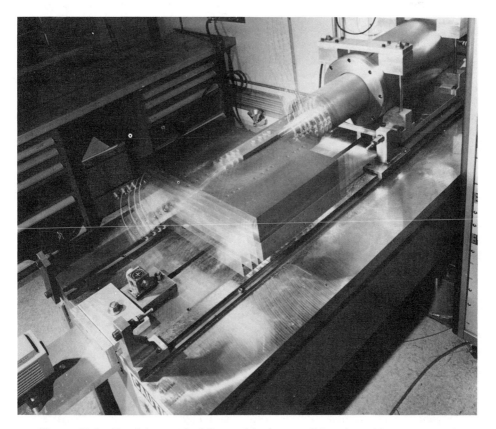

Figure 7.1   Precision test bed for positioning an air bearing table (courtesy of Martin Marietta Energy Systems, Inc.).

bearing table along a single axis of travel. This table is supported on a set of vee and flat guide ways by the air bearings and its position along these ways is measured by a laser interferometer to a resolution of 2.5 nm (0.1 microinch). The actuator that is used to move the table is a dc linear motor that is described in Figure 7.2. The computer accepts the position set-point commands and the position feedback information (in the form of forward or reverse pulses) and generates a position error signal based on the difference between the actual and desired positions. This error signal is used by the motor drive circuitry to propel the table to the desired location. If the position commands are changed rapidly with respect to time then the table is unable to complete a given move command before another move command occurs. In this situation, the table moves with a velocity that is equivalent to the rate of the command pulses multiplied by the weight of these pulses, although it is always slightly behind the commanded position at any given instant in time (a characteristic of feedback control systems). Figure 7.3 shows the positioning error, or following error, that is calculated by the computer for different rates of changing set-point commands or velocities for the system shown in Figure 7.1. On a different scale, this same approach is utilized

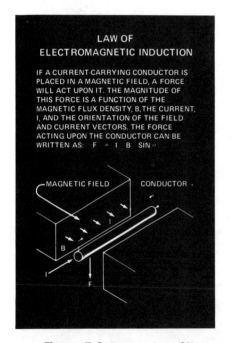
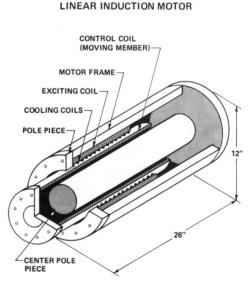

Figure 7.2 Description of linear motor operation (courtesy of Martin Marietta Energy Systems, Inc.)

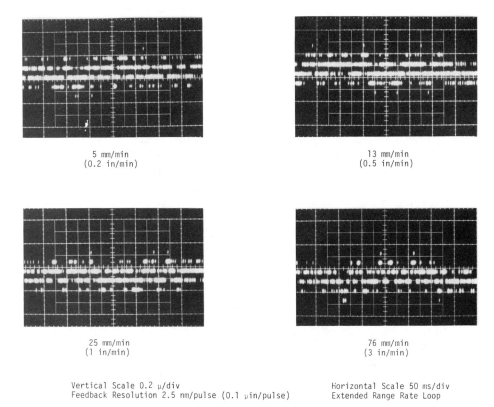

5 mm/min
(0.2 in/min)

13 mm/min
(0.5 in/min)

25 mm/min
(1 in/min)

76 mm/min
(3 in/min)

Vertical Scale 0.2 µ/div
Feedback Resolution 2.5 nm/pulse (0.1 µin/pulse)

Horizontal Scale 50 ms/div
Extended Range Rate Loop

Figure 7.3 Computer-generated following-error signal (courtesy of Martin Marietta Energy Systems, Inc.).

to control the positions of the axes on a modern machine tool (the command and feedback pulses are significantly less precise).

Programmable Logic Controllers

Programmable logic controllers (PLCs) are devices that simulate the actions of a digital logic network through the use of a digital computer as well as providing an option for use in data acquisition activities. The PLC is essentially composed of three main elements: the digital processor, the I/O circuitry and the software that enables the system to be programmed to perform the desired functions. Instead of connecting process signals to a relay network to generate the required control actions, these signals are input to a specially programmed computer for analysis and control. The PLCs internal program performs the same function as the old logic networks and outputs the appropriate discrete activation/deactivation signals to the devices that are being controlled. These PLCs can have a large impact in those applications that depend upon relay circuits to control an operation. The types of device control problems that are most frequently addressed are those involving Boolean algebra, for example, the on or off status of a particular function is dependent upon the on or off status of the appropriate inputs. For instance, a typical Boolean algebra task might be to obtain the necessary circuitry to control a pump in a particular manufacturing operation. A statement of the problem could be to insure that pump A is activated whenever switch A is in the manual on position or whenever switch A is in the automatic position and limit switch B is activated, and that pump A is always deactivated whenever switch A is in the off position.

Instead of wiring the hardware components together with a switching circuit, as was done in the past, the desired control relationship between the various entities can be programmed into the PLC computer (although interface circuitry is required to establish compatibility between the external devices and the computer). An additional advantage is that a detailed knowledge of switching circuits is not required to obtain a usable system and the computer's programming language is easy to understand even without a background in computer systems. Also, since a complex combination of relay coils and contacts is not used to resolve a controls problem, such as the one described above or the magnetics associated with a machine tool, the network design can be readily changed by modifying the computer program. This is much easier than rewiring the hardware network as was required in the past. This flexibility is also important when it is necessary to make changes on the shop floor. In addition, distributed PLC systems using multiple processors often allow the user to segment and partition the system in a more organized manner than if an attempt were made to employ a single, large PLC to control everything. Effective use of the distributed system approach leads to central monitoring and programming of a process.

With time, the functions of the PLC have expanded so that in certain respects they offer some of the features of the more complex numerical control (NC) systems such as fiber optic communications. In addition, these devices have been integrated into various categories of numerical control systems. This permits the NC system to offer the PLC features without the necessity of adding another external box.

Numerical Control

Numerical control (NC) technology began having an impact in the manufacturing world in the 1940s as an outgrowth of the production requirements for helicopter rotor blades. Since an air foil is a complex three dimensional surface it is impractical to attempt to fabricate it on a manual machine using only the operators ability to generate a curved surface. Initially, the NC manufacturing approach was limited to providing sets of coordinates that were used in a manual fashion to perform two axis machining operations so that the operator had some guidance on how to synchronize the machine's axes at discrete points. While this machining method was relatively successful, it was realized rather quickly that significant process improvements could be achieved if a servo mechanism could be constructed to replace the existing manual axis feed mechanisms (hand cranks). This would mean that it would no longer be necessary to depend on the skill of an operator to achieve the positioning synchronization of the machine's axes that is required for contour machining. Thus, the development of the technology required to continuously position a machine's axes to a series of predefined locations gave rise to todays multi-axes contouring machine tools.

Early NC systems emulated the technique of synchronizing the machine's axes at a series of points in space but did not necessarily provide this matching of paths between these points. This point-to-point operating technique was an improvement over the previous method but it did not provide a true continuous contouring capability. Today, NC systems are used to completely control the operation of a machine's axes based on a series of commands (called a part program) that reside on a punched tape or in a computer memory. During operation these commands are read from the appropriate part program storage device and used to generate the series of servo system commands needed by the axes drives in order to fabricate the desired workpiece. These part programs consist of tool path commands plus the necessary instructions (called m- and g-codes) to control spindles, coolant pumps, tool changers, and so on.

While it is desirable to maintain continuous synchronization of the machine's axes throughout the complete machining operation, it is impractical to store every position command increment in the part program since the program length would become prohibitive rather quickly. Instead, a prescribed contour is broken up into a series of straight line segments (as shown in Figure 7.4) that approximate the desired shape to the required degree of accuracy, and each of these straight line segments becomes a command or part program block to the NC. This is similar to the series of axes synchronization points that were used in the original manual NC activities. This significantly reduces the length of the part program, although in some applications the tape may still be hundreds of feet long.

The tool-path commands consist of a series of distance moves and the desired velocities. Through an internal process called interpolation, the NC generates the incremental axis commands required to execute the part program block while keeping the machine's axes synchronized at all times, assuming the servo system can keep up with the positioning commands. Unfortunately, during high speed machining the servo system may encounter difficulties in maintaining the axes synchronization. As mentioned earlier in this chapter, a closed-loop servo system typically lags behind the commanded position by an amount that is proportional to the rate at which the position commands occur. During straight-line moves

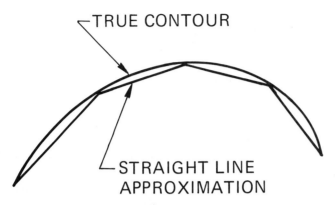

Figure 7.4   Use of linear interpolation to approximate a curve (courtesy of Martin Marietta Energy Systems, Inc.).

this is no problem because the axes maintain their synchronization even though they are lagging behind the commanded position. In addition, for gradual contours, the change in axes speed from one block to the next is so gradual that the tool path is still acceptable. However, during high-speed machining of corners this factor becomes a problem so that it is necessary to make programming adjustments to avoid errors in the tool path.

The commands for the part program tape are generated ahead of time using manual calculations for relatively simple shapes. For more complex contours, an off-line computer program is utilized that analyzes the desired part shape and calculates the straight-line segments required to approximate this shape to the necessary level of accuracy. At the time of use, the resulting part program commands are decoded and processed in the NC by dedicated electrical circuit that perform the necessary machine control tasks (interpolation, axis position control, spindle speed, etc.).

Initially, the NC systems utilized vacuum tubes and relays to accomplish their tasks of controlling the machine's axes and the necessary peripheral equipment. However, over the years these systems have followed the evolution of the computer, moving on to transistor and integrated circuit technology for the internal data manipulations. Today, NC systems with extensive dedicated circuitry no longer are being manufactured. Instead, the systems are designed around internal computers and called numerical control systems.

Computer Numerical Control

In comparison to the earlier NC units, computer numerical control (CNC) systems provide a previously unavailable degree of flexibility to control systems vendors and users. With the CNC unit, the manner in which the system functions as well as the purpose of the unit's switches and I/O ports is defined by the executive program that resides in the computer's memory. This feature makes it possible to alter the operation of the control system for changing applications by modifying the computer instructions that are contained within the executive program. Unfortunately, this was an awkward procedure with the early generations of CNCs because the vendor was the only practical source of a modified executive program. Vendors are understandably reluctant to make their proprietary software available to users since this represents a considerable investment of resources. In addition, even if this information is obtained it is a risky process to modify the executive program without an in-depth knowledge of the various interactions that occur within the program. However, a problem with depending on the vendor for program alterations was that frequently the turn around time was prohibitive or the vendor simply did not offer control systems expressly tailored to specific users.

Later models of CNC systems have alleviated the customization problem by permitting the user to have complete control over the machine's configuration (within the limits of available hardware). This flexibility has been achieved by designing specific set-up functions into the executive program that permit the user to configure the unit's operation through easily executed procedures. However, while these procedures are not complex, there is a level of protection built into the procedure so that someone cannot easily tamper with the system's set-up definition. Of course, if the system definition is altered after delivery by the vendor, it is the responsibility of the user to maintain the documentation required to define the current mode of operation of the control system. This flexibility does exist in conjunction with the potential for maintenance nightmares if faulty documentation of the system definition is permitted to occur. However, the need for accurate system documentation throughout a manufacturing facility is of paramount importance anyway so this is not really an additional burden. In addition, the ability to be able to do things such as reconfigure the function of an output terminal, change the resolution of the axes feedback signals, or execute a special user-defined subroutine that initiates different responses based on the results of a particular series of events, far overcomes the effort involved in updating the system documentation. In fact, the practice of maintaining an accurate description of the system's current configuration will have many side benefits as the advantages of this operating procedure are recognized and employed in other, sometimes unrelated, activities.

An advantage that CNC offers to the control systems builder is the standardization that occurs across the product line. Aside from hardware variations due to differences in factors such as numbers of axes, types of feedback transducers and I/O terminals, the basic control configuration is identical from one unit to the next with specific printed circuit boards being plugged into a common backplane network. In addition, the control system software is configured for a specific machine application by enabling the necessary portions of a generic executive program. Those features that are not needed remain dormant unless they are activated at a later date.

Another advantage that the CNC systems have over the NC systems is the inherently improved ability to perform automatic trouble shooting procedures. Fault detection with the NC units was limited to a few simple items like a tape reader error or a servo system malfunction. In addition, indicator lights were the only means of communciation with personnel attempting to trouble shoot a problem. Because the CNCs have internal computers they are able to perform a variety of real-time trouble shooting tasks as well as execute specific diagnostic programs on demand. Also, communications with an operator or maintenance personnel are greatly improved through text output to the control's display terminal. The available information includes error messages that isolate specific subsystems and help files that guide the trouble shooting activities.

The number of control features that are available with a CNC unit is also significantly greater than what was offered with the previous NC systems. These new features include items such as automatic axis calibration and backlash correction (which corrects repeatable errors that occur in the direction of the axis travel), axis alignment compensation, temperature growth compensation, enhanced tool offsets, repetitive pattern operations that repeat a cycle a programmable number of times, operation from a group of programs that are stored in memory, shop floor editing, and so on. The advantage of having these and other options is that it makes the CNC system easier to use and it enhances the quality of the manufacturing operations. In addition, the incorporation of a graphics keyboard provides features that were previously available only through off-line computer systems. Automatic three-dimensional plotting of tool motions permits part program trouble shooting without having to machine a part. Also, graphics part programming allows the user to roughly sketch a part on the system's display screen and then enter the desired part dimensions from the part print. Once the necessary dimensions have been entered the CNC automatically creates a formatted part program.

A good example of the type of flexibility that is added to a machine through the judicious use of computers is the computer-controlled coordinate measuring machine (CMM). These inspection machines utilize servo-controlled axes and a self-calibrating probing system to measure the dimensions of parts of various sizes and shapes under the direction of a computer. This computer performs both the machine control task needed to perform the inspection operation and the data analysis task needed to compensate and display the inspection information in a meaningful fashion. Examples of typical parts that can be inspected include automotive engine combustion chambers, helical gears and turbine blades. In operation, these machines utilize automatic probe calibration techniques to define an inspection machine coordinate system. Then, automatic probing routines are employed to locate the datum surfaces that are appropriate for each particular workpiece, and to reference these datums to the machine coordinate system. Subsequently, the system automatically references the workpiece contour measurements to the datum surfaces through computer manipulations that account for the position and orientation of the part in the gage's work zone. At the completion of the measurement cycle, the computer analyzes the data and displays workpiece measurement information such as nominal dimensions, tolerances, scale factors, histograms, and so on.

Figure 7.5 shows an example of a CMM and its associated computer control equipment. This in-process inspection system is used to determine if the desired final part configuration exists within a rough forged machining blank. The intent of this pre-machining inspection operation is to permit the utilization of a machining blank that is as thin as possible. A relatively small amount of machining-blank excess stock is desirable because it minimizes the amount of time that must be spent in the machining cycle. At the same time, it is necessary to leave enough excess material so that the normal size and shape variations in the forming process

Figure 7.5   Coordinate measuring machine for inspecting machining blanks (courtesy of Martin Marietta Energy Systems, Inc.).

will not produce an unusable blank. From a manufacturing quality control and productivity standpoint it is necessary to know that the desired part is attainable from a particular blank prior to beginning machining operations, otherwise the machining time may be wasted if the blank does not clean up by the time the finish workpiece dimensions are reached. (The alternative is to form a blank with so much excess stock that success is assured, but this approach also guarantees that machining resources will be wasted.)

Following the forging stages, but before any machining operations are attempted, the CMM system automatically inspects the rough part and transmits the data to a computer for analysis. The system software attempts to fit the dimensions of the finished workpiece within the available material. This data analysis step determines if sufficient excess stock exists to obtain the desired workpiece and provides product quality feedback information to the forming operation. In addition, if the blank should be aligned, on the machine tool, in a particular way to provide the most leeway during the machining cycle, then this information can be provided to the operator who will fabricate the finished workpiece.

Another feature of many CMMs, which is shared with a number of robot control systems, is the ability to learn a series of motions for future use. In this instance, an operator uses a joy stick to move the machine axes through the necessary sequence of motions that are required to accomplish a particular task. The control computer records these motions and converts them into an internal command program that can be recalled for later use. Unfortunately, the computer-assisted programming techniques available for use with robots and CMMs, for generating machine commands from workpiece dimensions, are not as sophisticated as the methods available for use with machine tools.

Whatever method is utilized to generate the information that defines the machine motions, a variety of mechanisms are available for loading the program into a stand-alone, computer-controlled machine. These techniques include I/O devices such as a tape reader, disc drive or manual data entry keyboard. In addition, the connection of a group of discrete CNCs to a supervisory computer, or host, permits data to be transmitted bidirectionally over a data link without requiring a tape reader. This type of configuration is defined as a direct numerical control system.

Direct Numerical Control

Individual Machine Control Systems.   Direct numerical control (DNC) is the next level up in the increasing chain of NC sophistication that begins with the now obsolete hardwired NC unit, and is followed by the CNC. The stand-alone CNC unit is computer controlled but it must get its operating instructions from an external source. This limitation requires the part program to be loaded manually into the CNC memory through a keyboard, tape reader, floppy disc, and so on. In contrast, the DNC system has a direct link to the source of the part programs and does not require manual intervention to physically transport the operating instructions to the internal memory of the individual machine control systems (CNC units). DNC systems consist of a local machine controller, the necessary hardware and software for communication with a higher level computer system and a host computer that is capable of meeting the needs of a number of machine control units. The local system may be an NC unit that has been retrofitted with peripheral memory or it may be a complete CNC. (At times, NCs, NCs with memory and CNCs are all termed NC in the jargon of the machine-controller industry.) The hardware and software that facilitate communications between the host computer and the machine control system may reside within a CNC unit or a local minicomputer may be used for this task.

One of the important requirements for a successful DNC system is that the individual machine control units have sufficient internal resources to continue operation for a lengthy time period, in a stand-alone fashion, if the information flow to and from the next higher level of the system hierarchy is interrupted. In an elementary DNC system, the only information flow is the handshaking communications needed to request a series of operating commands (the part program) and the actual part program data. In a more sophisticated system, a wide variety of additional information is transmitted in both directions. This data flow can consist of items such as station messages (mail), monitored data (status), CNC system executive programs, operational procedures (including graphics) and analysis results (including process offsets).

In addition to serving a number of machine controllers, the host computer may also function in other ways that are related to more generic computing tasks such as scheduling, planning or general data processing. The main constraint is usually the allowable delay that can be tolerated, while waiting for data, when it has been requested by one or more local machine control systems. It is undesirable to wait a long time when a new part program needs to be downloaded into the local computer memory because this is lost machine time as well as a disruption to the normal flow of the process. This delay also can become extended further as operators become occupied with other activities. In the event of an extended loss of communications with the host computer, the local minicomputer associated with the individual CNC system is usually capable of providing the machine control unit with needed part program information as long as the information has been previously downloaded. Also, the control system tape reader and manual data input keyboard provide an additional backup capability for data entry. However, any processing data that is usually monitored and sent to the host will be lost unless a special provision is made for its temporary storage on a local peripheral device.

As mentioned above, the DNC host computer system maintains a data base of part programs that may be selected for use on a particular machine control system. This can easily be an extensive amount of information that is sensitive from a quality standpoint, as well as being the result of the expenditure of significant company resources. Therefore, it is imperative that the number of individuals who are permitted access to this data for editing purposes be limited to as small a number as possible without significantly hindering the turn around time required to obtain the inevitable program modifications. In addition, there may be some confidential data that requires a restricted access procedure, such as a password verification step, even for reading the stored information. Also, updates to the data base must be recorded and a complete history of system access should be maintained along with adequate backup copies of the data base. This provision assures that operations can be easily restored following an unexpected incident that results in the corruption of data files. While this may seem to be an unlikely event in some instances, it is very unusual to have a computer system that never experiences a problem that results in the loss of some information.

Figure 7.6 shows an example of one type of hardware configuration that can be used in a DNC system for machine tools [1]. This system utilizes a host mainframe computer that executes a DNC supervisory program that is able to support the needs of each of the lower level control units through serial data

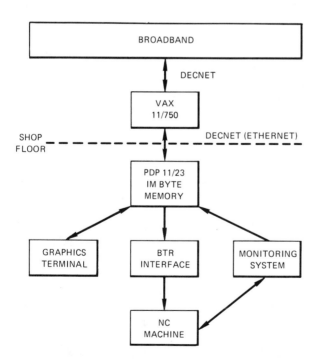

Figure 7.6 Hardware configure for a typical DNC system (courtesy of Martin Marietta Energy Systems, Inc.).

communications. A subsidiary DNC program operates on the shop floor computer. This shop floor computer system is capable of automatically storing and retrieving part programs, in response to an operator's request, through interactions with a data base on the host computer. In addition, this system is able to monitor the appropriate machine parameters and send this information to a data base on the host system, as well as coordinate communications between the shop floor and host computers and interact with the host computer mail system. The shop floor processor directs the information flow between the host computer and the machine control system and has one megabyte of memory which is dedicated primarily to part program storage. Also, this processor is interfaced to a graphics terminal so that machining procedures, complete with graphics, can be presented to the machine operator. In addition, the same network can be used to send digitized pictures (digital television) to the shop floor. This visual information provides a clear description of how different operations should be performed. These pictures can show step by step how critical aspects of a task are to be carried out so that there is a reduced chance of problems due to a lack of the proper communication of the required manufacturing procedures.

One advantage of the DNC system is that part program information is loaded directly into the machine control system's part program storage memory without depending upon the integrity of the paper tape reader. A behind-the-tape-reader (BTR) interface or other serial input port is employed to load the machine commands from either the host computer or the shop-floor computer, just as if the tape reader were in operation. In fact, in the case of a BTR interface, the control system doesn't detect the difference because a switch is utilized to temporarily connect the serial data output from the data-source computer to the tape reader interface (in place of the usual output from the tape reader read-head).

Whatever data transmission technique is used to bypass the tape reader, the result is that the frequent maintenance problems associated with correcting the cause of tape reader errors is avoided. In addition, the availability of up-to-date part programs is improved since these programs also can be prepared on the host computer. Also, the time required to load the part programs into the control system memory is decreased because no manual intervention is required to carry the tape to the machine, load it onto a tape reader reel and activate the tape reader. Plus, if part program preparation is performed on the host computer system (or another computer with which the host can communicate), then a significant reduction in the time for part program editing can be achieved since the editing operations can be implemented from the shop floor. (This is not necessarily a desirable mode of operation in all cases, due to work scheduling conflicts, but it can be a useful option in many instances.)

Some other advantages of a DNC system are related to its ability to perform automatic process monitoring and data analysis. Obviously, anything about a DNC-controlled manufacturing process that can be monitored also can be accumulated in a data base. Therefore, it is relatively straightforward to incorporate statistical process control techniques into the system's operations, although the decision on what are the most applicable process parameters for monitoring may be debatable. (The previous chapters on deterministic manufacturing and process monitoring describe in more detail some of the options that are available for manufacturing control with a DNC system.) Status monitoring also is useful in evaluating system utilization as well as for recording the sequence of events that has taken place when there is an unexpected occurrence. Often, it is difficult to reconstruct the events that happen prior to a real or apparent system malfunction. This problem is frequently magnified when the only observer is the system operator who may not even remember exactly what happened due to the excitement generated by the event. Also, there are some characteristics of a malfunction that the operator has no method of detecting since these phenomenon occur within the machine's different subsystems. The result is that the operator is able to detect only the effect of the malfunction, but not the cause of the unwanted action. In addition, historical status information provides valuable guidance in determining if the anomoly is due to the manufacturing equipment or the manner in which it was used.

Cell Control Systems.

A manufacturing cell is a group of machines that act together in a coordinated fashion to perform a specific task. Usually a part is processed at multiple stations within the manufacturing cell and the transportation between these stations occurs through the use of an automatic material handling system. In a machining environment, a typical cell might consist of one or more vertical machining centers with automatic tool changers and a pallet transportation system to support workpiece/fixture and cutting tool exchange. In addition, broken tool monitoring and automatic setup features may be included to provide continuous unmanned operation. The task of the computer-based cell control system is to coordinate the activities of the group of machine tools, robots and other factory devices that are integrated together into a manufacturing unit. Since the cell controller is needed to synchronize the activities of the various cell elements, it should not be built into a remote host computer. The cell controller should be a separate entity that is located near the cell and is unaffected by problems with a DNC host computer.

In its simplest form, the cell controller is nothing more than an equipment sequencer, such as a PLC, that controls the order in which actions are carried out within a cell. However, these systems do have limited data storage, as well as relatively elementary computational and communication capabilities. Specialized cell control systems are able to support other specific tasks that are desirable for a cell controller. These features include communications with other devices outside the cell, monitoring and controlling the operations of the devices within the cell, and overseeing cell functions such as tool management and report generation. Of course, these systems are also more expensive than the PLC-based units.

It should be apparent that in a DNC manufacturing cell the boundaries between the operations that are most suitably performed by the different computers are somewhat tenuous because of the overlap in capabilities. Also, the computer network architecture is different than what might be used in a conventional DNC system where parts are not automatically transported between the different stations. A cell controller is not needed with the conventional DNC system. However, there is some overlap in capabilities between the cell controller and the DNC host computer in a DNC manufacturing cell. In the DNC cell, it may be acceptable to utilize a relatively unsophisticated cell controller to accomplish the synchronization of operations and depend on the DNC system to perform the other tasks.

Computer-Aided Manufacturing

Computer-aided manufacturing (CAM), computer-aided design (CAD), computer-aided engineering (CAE) and computer-aided whatever (CAX) are all acronyms for operations in which computers are used to enhance the quality and productivity of the particular processes. Computer-aided manufacturing encompasses automation, control, planning, preparation of machine commands and the monitoring of the manufacturing operations. However, there is not a distinct delineation between these various disciplines due to the evolutionary processes that are taking place. Because of this territorial overlap it is often a waste of time to attempt to classify a topic as belonging strictly in one area or another. However, this section will be limited to a discussion of CAM since it is the topic that is most closely related to the overall subject matter of this text. Nevertheless, some crossing of boundaries into the other generic catagories will occur, but this should not be confusing if excessive attention is not directed toward the subject classification.

Figure 7.7 shows an example of the operational steps that are followed in most manufacturing facilities to fabricate machined workpieces. Initially, a design is acquired from the customer in the form of blueprints and/or other product specification material. Then, the processing steps required to successfully complete the job are defined or planned. Of course, the sophistication of the resulting process plan depends on the complexity of the desired workpiece and the similarity of this job to other tasks that the facility produces on a routine basis. However, this process plan may be thought of as a blueprint for the entire manufacturing process. In addition, this plan can be structured as rigidly as the documentation produced by a computer-aided process planning (CAPP) system or as informal as a sequence of steps that an individual organizes in his or her mind.

In the next phase of the production process, the generation of specific shop-floor operating procedures, NC tapes for machining and inspection operations, and tooling designs occurs. When performed without computer assistance, these manual planning activities are time-consuming and expensive. However, the

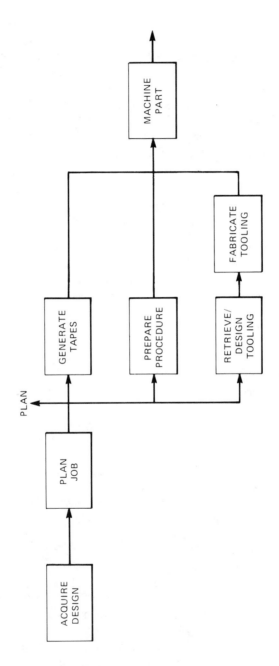

Figure 7.7 Steps required to fabricate workpieces (courtesy of Martin Marietta Energy Systems, Inc.).

alternative exists to employ automated operations planning for parts that fall within the same part family. To accomplish this, the planning logic can be formalized and converted into computer code. Then, the planning information can be automatically generated when the necessary part-specific information is provided to the computer planning system. In a broader scope, computer-aided planning and scheduling activities can be expanded to include the automatic allocation of all of the operations performed in a manufacturing facility. This expansion of activities leads to the topic CIM which will be addressed in the next major section of this chapter.

Figure 7.8 shows an example of a CAM system that operates with the DNC hardware shown earlier in Figure 7.6 to manufacture hemispherical workpieces [1]. This system utilizes a CAD-generated product definition that resides in a computer data base to automatically create the information that is necessary to fabricate a specific product. For those workpieces that are members of a part family that has been previously defined in the computer data base, the system is able to interactively request the needed part-specific input information that is required to specify the necessary processing requirements. Then, the system is able to automatically create a process plan and tape images for machining the part, as well as the text and graphics for the operating procedures.

As the new workpiece design undergoes automatic processing, the generative process planning (GPP) system receives the geometric description of the product plus the pertinent specific manufacturing requirements. Then, the system utilizes the previously defined standard manufacturing logic to generate the required instructions for fabricating the part. The resulting NC information is in the form of an automatic program tool (APT) source file. In most instances, this data must be compiled and processed again by a special computer program called a postprocessor before it can be used with a specific machine-controller combination.

The quality of the tape image that is formulated by the postprocessor (the part program that will be used to machine the workpiece), is checked automatically before it is used in the actual production operation. A machine simulator executes a mathematical model of the machine/controller pair to verify that the tape image will result in the desired workpiece. Following acceptance of the tape image, the procedures and part programs are loaded into a data base where they may be accessed from the shop floor at the appropriate time.

Another example of an operating CAM system is the AT&T manufacturing facility located at the AT&T Technologies Works in Richmond, Virginia [2]. This facility is operated as a job shop and is controlled by a totally integrated computer system. In this instance, the application is the high volume manufacturing of printed wiring boards (PWB). In the overall manufacturing cycle, the PWB is usually the last component designed, the first one needed for assembly and the one that is changed the most frequently. In addition, the facility has over a thousand active PWB codes at any given time, with an averge life expectancy of less than two years. The result would be chaos without the assistance of a

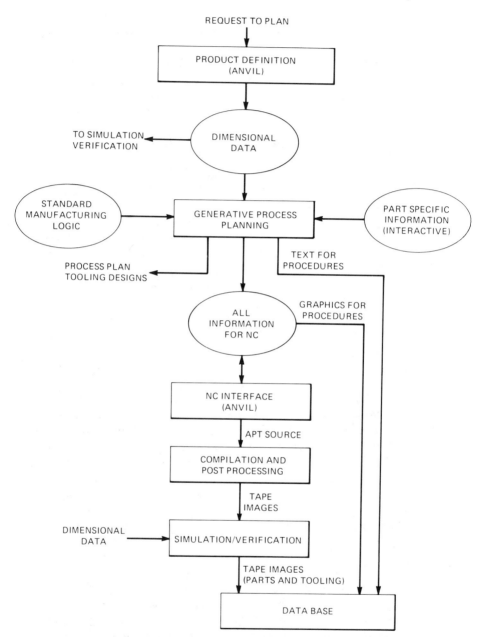

Figure 7.8 IMICS manufacturing information generation system (courtesy of Martin Marietta Energy Systems, Inc.).

computer system to manage the massive amount of information that is required to fabricate the products in an effective manner. The solution to this production task is a flexible, fast reacting CAM system that is able to meet the data handling and control requirements of the manufacturing process.

At the beginning of the manufacturing cycle, new PWB designs are teleprocessed in digital form from the various remote Bell Laboratories design locations. The rapid transfer of this data is accomplished through the use of a corporate wideband transmission network. This transmitted information includes a definition of the circuit board's electrical interconnections, plus the locations of the drilled holes and the necessary test points. When the new design is received at the Richmond Data Center, its features are compared automatically with the previously defined families of PWBs. When the correct match is found, the system automatically generates the NC files for machining and testing the boards as well as plots showing the resulting machining/testing operations. Following the review of this information by an engineer, the data required for production is released to the manufacturing areas.

During PWB production, shop status information is provided by maintaining a queue for each machine and recording the completion of each lot via a bar code reader or keyboard. When the operator selects the next job from the queue, the processing instructions are automatically printed out at the machine. This insures that the operating procedures are always up to date. In addition, the transaction history provides process management information that can be utilized for determining process yields, individual machine throughput, manufacturing efficiencies, and so on.

The previous example discussed a CAM application in a fabrication process that required many different designs and fabricated products in a given year's time. In addition, CAM techniques also are useful in "few of a kind" applications. One example of this type of situation is the Ingersol Milling Machine Company's computer-directed manufacturing process for the fabrication of automatic machining centers [3]. Ingersol manufactures approximately 50 of this type of machine tool each year and all of the systems are essentially special orders. This means that the average lot size at this facility is between two to three units, half of which are new designs and half of which will never be used again.

Ingersol utilizes an integrated CAD/CAM system which includes application programs for graphics construction, display, and plotting, in addition to a graphics data base management system. The initial phase of the manufacturing process begins with the creation of an electronic design that is composed of conventional two-dimensional views that are linked together using computer software. This design process produces a geometric model that is stored in a computer and used to produce assembly and piece-part drawings. Using this geometric model, the NC programmer plans the machining methods, establishes the machine setups, and selects the necessary tooling. Then, a light pen and digitizing tablet are used to indicate to the computer program which surfaces are to be cut as well as the desired tool path. At this point, another computer program is utilized to generate the NC instructions and test the final part program for errors that could result in collisions on the machine tool.

Computer-Integrated Manufacturing

The scope of computer-integrated manufacturing (CIM) includes those topics covered in CAM, CAD and CAE as well as additional, sometimes neglected, manufacturing activities such as scheduling, communications, inventory, shipping and receiving, and so on. Computer-integrated manufacturing activities are intimately involved in the gathering, manipulation and distribution of process and management information, in a timely manner, throughout a manufacturing operation. At present, there are many definitions of CIM. In some instances, almost any manufacturing activity that involves a computer is termed CIM. However, in other situations, the computer system must be fairly complex before it seems to be worthy of the CIM title. Regardless of this discrepancy, most CIM activities are still involved with solving the problems associated with the various pieces of the overall technology puzzle. These areas of activity include DNC systems, automatic process planning, automatic guided vehicles, computer-controlled scheduling, and so on. This approach in dealing with the subsets of the overall system is a practical method of arriving at an eventual solution in a reasonable manner (as opposed to attempting to design everything together into one large system before the details of the process elements have been reasonably well defined). This is a good technique to pursue in solving the problem as long as the communications issue is not neglected while attention is focused on the individual technical areas. Attempting to implement computer integration throughout all the various phases of a manufacturing operation can lead to immense problems unless the different systems are able to exchange useful information in a meaningful and timely fashion.

The long term goal of CIM activities is to develop a system that is capable of integrating all the functions of an entire manufacturing operation. While this would be an extremely complicated subject to address, it is possible to discuss the basic concepts that are involved by examining a small portion of the overall system. One good example for this purpose is the topic of cutting tool management. The information that is required to purchase the necessary cutting tools for different manufacturing processes includes a list of acceptable vendors, pricing information, part numbers, cutting tool distributors, alternate sources of material, and so on. In addition, specific cutting tools must be associated with a particular fabrication process. This requirement means that the tools must be linked to certain production part numbers, as well as specific machines and types of operations. Also, the individual tools (and families of tools) must be monitored to establish "end of useful life" limits, checked for rework potential and whenever applicable, followed through any regrind cycles. A computer system which maintains all of this process information in a data base permits rapid evaluation of the tooling operation. In addition, it may provide auxiliary spinoff information that would otherwise be unrecognized, such as the need for a tooling change or overhaul of a particular machine tool.

In recent years, the complexity of manufacturing tooling has increased significantly. As a result, it is impractical as well as excessively expensive to maintain a miniature tool crib at each work station. However, it also is very nonproductive to have machine operators spend a lot of time waiting while tool kits are assembled for a particular job. A solution to this dilemma is to utilize a computer-based tool management system that provides the required material-management technology. This type of system is capable of controlling the supply of a large number of cutting tools without inflicting excessive delays or oversized tool inventories. This may be accomplished automatically when the tool management system is a portion of a larger facility management system. In this case, the information which prescibes the tools needed on a particular day is available automatically, through the planning and scheduling operations, with sufficient lead time to permit tool room personnel to have the necessary tool kit available when it is needed. In addition, transaction logs may be used to record who receives the tools as well as what operations are utilizing these tools. This information can be assembled in a data base to obtain a complete history of each tool's usage. Then, this data can be utilized in studying the characteristics and/or eccentricities of different processes.

From the broader manufacturing viewpoint, the difficulties associated with efficiently controlling a complete manufacturing facility are similar to the previously discussed tool management task. However, the scale of operations is expanded quite significantly. While the problems associated with a complete manufacturing complex are magnified in terms of both size and complexity, in relation to the tool management task, there is still a large degree of commonality between the two situations. Both tasks are involved with ordering materials to meet a specific manufacturing schedule while avoiding excessive inventories, scheduling work through the necessary processing steps in a productive manner, fabricating products to meet specific tolerance requirements, and shipping the products in a timely manner to meet the needs of the customer. However, because of the increased complexity of the task associated with controlling a complete manufacturing facility, the problem of obtaining an accurate assessment of the status of each phase of the individual operations is much more crucial. Accurate up-to-date information on the status of each subsystem within the manufacturing facility is required to enable the accurate operations analysis that is required for effective work scheduling. In addition, the hierarchy of computer-controlled functions must operate as a distributed control system so that each subsystem is relatively self-sufficient. This is necessary to provide the manufacturing system with the ability to continue operations for an extended time period in the event that problems develop at a critical location within the network of computer systems.

Knowledge-Based Systems

Knowledge-based systems (KBS) depend upon expertise that is built into computer programs. However, these programs are drastically different from conventional algorithmic programs. Traditional computer programs work on data to produce information through the use of algorithms that precisely describe the way the data must be manipulated. A shortcoming of this approach is that these programs cannot deal with uncertain or missing data, nor can they explain the significance of the information that is generated. To convert information into knowledge requires intelligence. Knowledge systems combine information about a subject with different ways to think about the information.

Knowledge-based systems is one of those high tech items that mean many things to different people. In some instances, KBS may be used to refer to a technique for recognizing "big ones " or "little ones" that occur as anomalies in a product stream, while in other cases, the KBS may be employed to recognize characteristics such as the identity and orientation of an object and to make processing decisions based on this information. For the purposes of this discussion, KBS will be considered to be an application of that elusive, ever-changing concept called *artificial intelligence* (AI) in which a computer system executes a program that simulates the response that might be obtained from a human expert.

Beginning in the late 1970s, AI received a lot of promotion from leading academic institutions. In addition, the Defense Advanced Research Projects Agency (DARPA) created a strategic computing program to take advantage of the potential dividends [4]. As might be expected, the AI focus has evolved with time and some of the originally conceived projects are still too complex to be solved within the present capabilities of existing AI systems. At the same time, other projects have been identified in which it has been possible to usefully apply AI techniques to solve significant problems. At present, the near term applications for AI are limited by the available technology. However, the long term prospects appear to be promising.

Two general categories of KBS activities are those applications which attempt to duplicate natural human abilities (vision, language processing, etc.) and those which attempt to duplicate learned skills or expertise [5]. The first category has a benefit in manufacturing processes of providing improved machine control. The second type of application is commonly referred to as an expert system (ES). This classification of KBS is concerned with the automation of tasks that are normally performed by specially trained people. Examples include systems that are designed to improve the capabilities of the experts themselves, to capture the knowledge of personnel who are retiring and to provide the capability to quickly transfer critical skills and knowledge throughout an organization.

To date, other systematic approaches, that are based on computer systems, such as group technology, decision support and management information systems, also have achieved some successes in the areas addressed by ES. However, the key difference between these technologies and ES is the ability of ES to address problems that do not have a well-defined model or algorithmic solution. A second major difference is in the ability of ES to not only use a rule base for reasoning within the problem domain, but also to reason about its own inferencing process and provide rationale and justification for the conclusions that it reaches. The third difference is that in the construction of an ES the developer attempts to capture and represent the decision-maker's knowledge. Thus, building an ES is primarily a task of constructing a knowledge base and therefore embodies the notion of imitating the manner in which the human expert solves a problem.

There are five characteristics of those tasks that may benefit from a problem solution that employs ES. First, the existence of a human expert is crucial. In spite of science fiction stories about computer systems that rule the world, current ES technology is restricted to problem domains that require human expertise. Second, the knowledge base should be reasonably bounded and preferably domain specific. Problems which require common sense are not suited to an ES solution, but resource management within a small segment of a factory (such as a machining cell) where the system utilizes knowledge of the machines, tools, products and worker characteristics to maximize performance is a good candidate. Third, the performance of an expert on the particular problem should be significantly better than a beginner or apprentice. Obviously, developing an ES would be a waste of time if anybody could accomplish the task. Fourth, the problem decisions should require consideration of a variety of alternatives that may involve some uncertainty. The most suitable problems for an ES are those in which even a human expert may not be immediately aware of all of the possible alternatives due to the magnitude of the potential number of approaches. Fifth, the number of entities, their relative attributes and the useful relationships which define the solution alternatives or the decision factors must be bounded. One way to describe this aspect of the potential ES system is to require that the problem be characterized by individual independent tasks which generally require less than 5 to 10 minutes of consideration by an expert.

For those tasks that meet the above criteria, an ES is constructed by interviewing a number of experts in a particular field and documenting how they solve a certain group of problems. When this intelligence is stored in a computer data base, the appropriate facts can be retrieved at a later date to solve a problem that falls within the predefined area of expertise. The end result is that the computer system arrives at the appropriate intelligent action that meets the needs of a particular situation. This solution is selected by the computer software through a reasoning process that is based on the information in the computer data base. In addition, the ability can exist to modify and update the knowledge base in response to observed results from the operating environment so that actions are based on results, not just predefined facts.

The inherent benefits associated with KBS include the combination, distribution and preservation of knowledge. Human experts may not be readily available when they are needed for a given situation, and textbook and instruction manuals rarely present the full detail that an expert in a given field brings to a problem solving task. In addition, the successful combination of several individuals' expertise into one knowledge system (such as a medical diagnosis system) results in an overall package that is much more powerful than any single expert due to the inability of one person to know everything.

To date, the primary applications for KBS have been in medical diagnostics, and the utilization of KBS in the manufacturing environment may seem to be superfluous. However, useful applications range from maintenance diagnostics to the implementation of deterministic manufacturing systems. KBS is beginning to solve problems in the area of maintenance for aircraft [6] and manufacturing systems. Ford Motor Company utilizes a diagnostic ES to assist the maintenance crews at the Essex and Sterling Heights plants [7]. In addition, Ford is applying this technology in the development of a Service Bay Diagnostic System that provides an "expert in a box" to provide assistance to automotive mechanics. This "technical reference source" will be an invaluable information resource as automobile systems become increasingly complex.

Another example of the use of an ES for dealing with a complicated problem that has many potential approaches is the design of complex CNC systems. In this application, control systems are designed and tested on a computer system so that potential customers can participate in the design. Through computer simulation of the actual hardware the customer is able to tailor the system to his specific needs and observe what the implications are for various system designs. Another example of the use of ES in a manufacturing environment is in operations scheduling and product routing. Controlling an automated materials handling system can become an extremely complex task in a large manufacturing facility. One typical problem that can occur is for some work stations to become overloaded while others are without work. A successfully integrated ES will be continually aware of the status of the entire operation and be able to coordinate the various manufacturing activities to maximize the use of the available resources. In addition, the use of computer systems to implement the deterministic manufacturing philosophy, discussed in a earlier chapter, is another example of the use of KBS. Deterministic manufacturing operations automatically monitor appropriate process parameters and initiate the correct response from a statistical standpoint.

References

1. R. E. Neal et al., Integrated Manufacturing Information and Control System, Martin Marietta Energy Systems, Y/DX-659.

2. L. W. Phillips and W. L. Ogburn, The Evolution of Computer Integrated Manufacturing at AT&T Technologies' Works - Richmond, Virginia, Proceedings of CIMCOM '85, Anaheim, CA, (1985).

3. N. E. Ryan, CAD/CAM integration yields quality control, *Manufacturing Engineering, 98*:2 (1987).

4. G. R. Martins, AI: the technology that wasn't, *Defense Electronics, 18*:12 (1986).

5. R. J. Mayer et al., Artificial Intelligence—Applications in Manufacturing, Presented at AUTOFACT 6 Conference, Anaheim, CA, (1984).

6. D. Papenhausen, B-1B maintenance diagnostics gain a long memory, *Defense Electronics*, 18:9 (1986).

7. H. Gorman, The experts are coming, *Production Engineering, 34*:8 (1987).

CHAPTER 8

Computer Networks

Introduction

A manufacturing facility is likely to utilize multiple computers for several reasons. The first reason is to handle the required volume of data processing within a reasonable amount of time, while also providing a backup resource in the event that one system is unable to supply the needed computing capacity. Other reasons are to provide computer service to a variety of organizational entities that may be widely dispersed or because of the gradual manner in which computing systems are introduced to different areas of a factory. In order to obtain the maximum utilization to these diverse computing systems, it is necessary that they be able to communicate readily with each other. Data communications between different computer systems can occur in a variety of means such as mailing the information to another person who performs the data entry task, using a telephone to talk to this other person, physically transporting a computer-readable data record or utilizing a direct electronic communications link between them. This last communications technique is the one choosen when attempting to avoid time delays, although it is not necessarily the best approach from a cost, capacity or accuracy standpoint. In comparison to a direct electronic communications link, a truck that is packed with computer discs can certainly transfer more data in a given amount of time, with greater accuracy between two computer systems that are one mile apart. However, the time delay associated with using the same transmission for smaller amounts of data or for significantly greater distances becomes prohibitive.

People use the telephone as a mechanism for electronic communications that permits the transmission of information via modulated sounds called words. In addition, there is a certain path of telephone poles and wires that these electrical signals follow to travel from one user to another. Electronic communications between computer systems occurs in a similar fashion through the use of a hardware interconnection called a computer network. The method or particular style that is utilized for this interconnection is called the computer network architecture.

The computer network is like a mesh or grid which interconnects multiple computer systems, over high speed data communications lines, and allows these systems to talk to each other. This information transfer means that the individual computer systems can interact cooperatively with each other to perform tasks that would be prohibitively complex for a single data processor. This information processing occurs so smoothly that most people are involved routinely with computer networks without ever realizing what is happening. Common examples include the electrical transfer of funds by the banking industry, airline ticketing operations and telephone switching systems. The purpose of each of these different networks is to allow a variety of data processing systems to share the overall job requirements through communications with each other. This system design, or architecture, means that one super computer doesn't have to be depended on to do everything. Instead, this approach has the advantage that various tasks can be allocated to a number of different computers so that the work load is distributed and the loss of one computer does not jeopardize the operation of the entire system.

The telephone system is a good example of a communications network. This system is designed to operate optimally with human speech (300-3,000 Hz), but it can also be utilized to transmit data between computer systems through the use of a hardware device called a modem. The modem translates the digital zeros and ones in the computer data into high or low tones that can be sent over the telephone link. At the other end of the system another modem translates the sound information back into digital characters that the computer system can understand. However, because the telephone system has a rather limited operating range, the data transmission is relatively slow with respect to the speed at which the internal data transfers occur within a computer. Ironically, within the telephone system, voice communciations are sometimes transmitted in digital form to minimize the impact of data transmission noise. This is accomplished by sampling the analog voice signal, transmitting the digital version, and reconstructing the voice signal at the receiving station.

In addition to modems and the wires that connect everything together, other types of hardware also are associated with computer networks. These devices include entities such as data processors, terminals, work stations, controllers and handlers, multiplexers, and concentrators. Most of these devices are concerned only with the transportation of information from one location to another. However, some of the elements such as terminals and data processors also may be found with a stand-alone computer system.

Figure 8.1 shows an example of a relatively simple computer network that is used by a manufacturer of snow skis to perform CAD/CAM activities [1]. This system utilizes a mainframe host computer to design and analyze the various characteristics of potential ski configurations. The work stations are utilized to perform tooling designs and to prepare NC programs to be used in the eventual

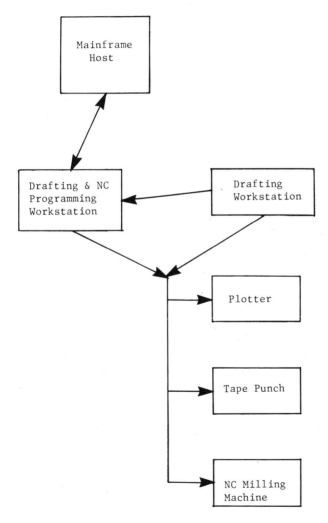

Figure 8.1 Example of a centralized control system.

hardware fabrication. These machining part programs can be down loaded directly to an NC milling machine, sent to a tape punch for hard copy generation or used with a plotter for tool path verification.

This chapter deals with computer networks from the viewpoint of a single manufacturing facility where information transfer is carried out between different on-site computer systems. This type of communications facility is called a local area network (LAN) and is characterized by relatively high data rates, a geographical area of coverage of about 1 kilometer and relatively low cost. Long distance networks between different geographical areas are not considered. The topics that are discussed in the following sections include the equipment used for the network data path hardware and the associated jargon, host and local computers, communications between these data processors, the International Standardization Organization (ISO) reference model, the General Motors spearheaded network protocol called MAP (Manufacturing Automation Protocol), and system reliability enhancements that provide continued service in the event that portions of the system are unavailable. Actual examples of the application of computer networks are presented rather than attempting to deal with the more technical aspects of this dynamic subject.

Data Path Hardware and Associated Jargon

Access to computer network facilities can be achieved through two different ways. One method utilizes devices called modems and permits the user to gain access to the system by employing the local telephone system. Modem is an acronym for a device that *mo*dulates and *dem*odulates digital information from a terminal or computer to an analog carrier signal that may be readily transmitted over a telephone channel. In addition, devices also are available to automatically place and answer telephone calls for the transmission of data. While this is a convenient means of establishing remote communications, it also may allow unauthorized users access to the system unless effective protection techniques are employed specifically to prevent this occurrence. (Data encryption also may be used to make information unintelligible to the unauthorized observer.) Although this access technique is flexible and cost-effective, there are some inherent disadvantages. The telephone system does not always use the same set of links to complete a call from one location to another so that the quality of the different circuits may be difficult to predict. In addition, extensive data communications activities such as might be associated with a time-sharing network will tie up a telephone system almost as effectively as a teenager. The other technique for accessing a computer network is to have a dedicated hard-wired connection to the system. This approach eliminates most of the problems associated with the telephone switching system, but it is more expensive since the user pays for the full-time use of the telephone lines. In addition, the flexibility is limited since the dedicated line only goes to the single termination point.

The various hardware elements that are connected to a computer network are often given the generic label of work stations. Examples of work stations that consist of conventional data processing equipment include devices such as computers, disc memories, printers, terminals, and so on. However, work stations also may consists of other types of equipment such as machine tools, robots and automatic guided vehicles. In a general sense, work stations may be described as the network equipment which corresponds to the human workers that occupy the various positions in an organization. The human work station has communications channels via a telephone or mail system plus procedural algorithms in the form of standard operating procedures and a job description. Temporary data storage is available in local memory while filing cabinets are used for long term storage and information backup. The network workstation utilizes metal or optical cables to transmit information throughout the network (satellite and radio frequency data transmission is available for long distance communications) and utilizes specific procedures in handling the data. In addition, several megabytes of high speed memory are frequently associated with the work station. Therefore, a convenient means of visualizing the operation of a network is to imagine a number of work stations that are interconnected so that they may function in a cooperative manner.

The data paths of a LAN consist of a variety of physical cables that are used to interconnect the various stations that are attached to the network. One type of cable used for communications is called twisted pair wiring. These cables have one or more pairs of wires that are twisted together along their length to provide improved protection from cross-talk or outside electrical interference. (Crosstalk is encountered when the signals on one set of wires are inductively coupled into other wires in a manner similar to the way a transformer works. The effects of this induced noise interference are errors in the transmitted data. This outside interference typically arises from electrical signals that are high voltage, high current or high frequency.) The twisted-pair communications medium is inexpensive and can be used with or without modems. However, the bandwidth is limited by factors such as transmission distance or system noise. Coaxial cable is a two-conductor wire cable in which the longitudinal axes are coincident. A noise shield is wrapped around the inner signal-carrying conductor so that it is relatively immune to electrical interference. This type of cable is widely used for communications because it is inexpensive and compatible with relatively high data rates, and it can be used in a broadband network. Optical fiber cable also provides a high bandwidth. In addition, it also has the advantages of being light weight, with a small cross section, which makes installation easier when small radius bends are needed. Fiber optics enhance data security because there is no significant electromagnetic field external to the cable and it is necessary to physically break a cable to tap into a signal. Another advantage associated with fiber optic cables is that they provide total electrical isolation and freedom from ground loops so that they are immune to electrical interference. At present, a disadvantage is the relatively high interface cost but this installation expense will be reduced in the future.

An alternative cable-system for intrabuilding data transmission is to utilize the existing power line wiring [2]. Utility companies have used carrier-current communication techniques over high-voltage lines since the early 1920s. University campus radio systems, intercoms, music systems, and burglar alarms are all examples of the use of this type of communications approach. Also, local area networks are commercially available that operate using power lines. This technique is feasible for use with local data communication rates in excess of 100 kbits/s and within certain constraints, data rates of 1 Mbits/s or higher are possible.

Wide band is another term that may be used to describe a broadband network. This terminology refers to the use of data transmission equipment that has a bandwidth (information-carrying capability) that is greater than what is available on those networks that are used for voice transmission. Baseband is a term used to describe the frequency band that information-bearing signals occupy before they are combined with a carrier in the modulation process. (The carrier is an analog signal that combines with an information-bearing signal in a frequency or amplitude modulation process to produce an output signal that is suitable for transmission over a particular channel.)

The physical data channel that is used in a communications network is an expensive resource, and most types of network peripherals only use a small portion of the channel bandwidth. In addition, those that do utilize the full bandwidth do not transmit and receive data continuously. Therefore, it is desirable to employ techniques for sharing the channel among multiple users. Three methods of accomplishing this goal are time division multiplexing, frequency division multiplexing and concentration (statistical multiplexing). Time division multiplexing is a technique in which the total available channel time is divided into slots or sub channels and allocated to the various users. This requires that the equipment that interfaces the various users to the network must store the low speed data in a buffer, transmit it in a high-speed burst at the proper time, and reconstruct the original low-speed message from the high-speed data. Frequency division multiplexing provides each user with continuous access to the physical channel by using individual frequency sub-channels (like television or radio channels). In this case, each subchannel functions as a separate lower speed channel. Statistical multiplexing is similar to time division multiplexing except that the time slots or subchannels are dynamically allocated. The advantage to this technique is that when a subchannel is not being used by one device it can be reassigned to another unit which has information that needs be transmitted.

Local and Host Computers

In its simplest form, a computer network consists of two or more computers that are able to transmit information back and forth over a communications link. A heirarchical network is one in which a host (supervisory) central processing unit (CPU) and one or more peripheral or local computers are interconnected as shown in Figure 8.2. In a distributed system, the local CPUs are capable of operating for an extended length of time without the necessity of communication with the host system. The length of time which the system can function in an autonomous fashion depends on the specific system design but it is generally related to the changing environmental conditions that the isolated system must accommodate. As long as things are relatively constant the local CPU may be able to operate indefinitely (assuming adequate memory resources), but as conditions change and new input is required from higher level sources, then access to the host becomes the limiting factor. However, alternative host sources may exist in some networks as well as the option of utilizing a manual method of information transfer. While available manual data transfer techniques are less convenient than down loading the data from a host system, in an emergency, these methods of data transmission provide a backup capability that can be utilized in the event that host communications are interrupted. An example is the ability to load a needed program from a backup tape memory so that the local computer system is totally independent of the host when necessa.y. For more complex systems, the processor that acts as a host to one system also may be operating under another host. This type of system configuration could have a shop-floor host that communicates downward with a number of shop-floor processors and upward with an area host, and so on.

Figure 8.3 shows the computer network that is utilized to support the data processing requirements for a two axis CNC lathe. This machine tool has been retrofitted with a tool changer and a DNC interface so the system design may be different than what might be utilized when performing an initial design. At the same time, retrofit installations are very common because it is rarely economical to rebuild a manufacturing facility from the ground up. This network utilizes one computer within the machine control system to control the basic machine tool functions, a second CPU to control the operations of the add-on tool changer and a third processor to act as a shop-floor interface to the DNC system host computer. This host computer executes a DNC software package that permits the bidirectional transfer of part program information as well as the accumulation of data that orginates on the shop floor. In addition, the DNC program is one of a number of concurrent tasks that are continually being processed by the host CPU. This means that the host's resources are allocated to the DNC system as needed, but this only requires intermittent communications between

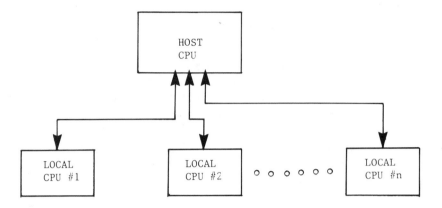

Figure 8.2 Relationship between host and local computer processing units (CPUs).

these two systems. At other times, the host is utilized to support some of the other data processing needs of the manufacturing facility.

Data transmission between the host and shop-floor CPUs shown in Figure 8.3, is accomplished using a serial data link that transmits information, one character at a time, at a speed of 9,600 baud. This baud rate refers to the number of signal level changes per second that occur on a communications line whether the changes are data bits or control bits. The data rate is the actual number of data bits that may be transmitted across the channel each second. (In general, this data rate need not be the same as or limited by the baud rate when phase modulation techniques are employed. With this approach, the phase of the carrier signal is used to transmit additional intelligence so that the data rate can exceed the limited bandwidth of the physical channel.)

Serial communication also is utilized to transfer the machining part program information between the shop-floor computer and the machine tool. This is done primarily because the data can be transmitted easily via the control system's serial tape reader input (through a BTR interface) and tape punch output. These two modes of data transmission allow part program information to be down loaded to the machine control unit from the host computer, as well as the up loading of shop-floor generated/edited programs for storage on the host system. Options for program execution include machine commands that have been down loaded from the host to the CNCs memory, execution of the part program from the shop-floor computer memory via the BTR interface and conventional use of the CNC system tape reader and part program memory.

In contrast, a parallel data path is available for the transmission of other information between points within the computer network. This data communication

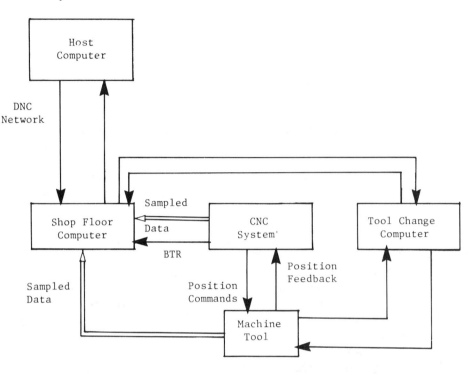

Figure 8.3 Computer control network for precision lathe.

occurs as 16 bit digital words as opposed to the single characters used with the
serial link. This method of information transfer is faster than the serial approach
but it requires a signal and ground wire for each bit, which can involve a lot
of wires in some instances. For the system shown in Figure 8.3, this data communication technique is used for transmitting the positions of the machine's axes
to the shop-floor processor. This approach also is utilized by the CNC to output
the m-codes that have been placed in the part program. (M-codes are inserted
in a part program to control miscellaneous peripheral functions such as turning
the machining coolant on or off, operating power clamps, etc.) This data is obtained from the parallel output ports on the CNC system.

The axes position information is needed for the in-process gaging and tool changing activities. The m-code information has standard functions within the CNC
system, but it also is needed in conjunction with the tool changer and DNC retrofit
system designs. Additional m-code operations were established that served to identify the in-process gaging data for correct storage in the data base and to initiate
specific phases of the tool change cycle. A parallel interface was the easiest way

to accomplish this retrofit task because it was only necessary to run the cables a relatively short distance. In addition, there were sufficient parallel data input ports available on the shop-floor computer to accept the signals.

Communications between the shop-floor computer and the tool changer computer are needed to synchronize the operation of the automatic tool changer and the machine's part program. Because the tool changer was added to the machine tool as a retrofit package, no direct data link exists between the tool changer processor and the machine control system computer. In addition, both of these computer systems are relatively limited in their ability to perform tasks beyond the control of their respective mechanisms. To achieve the needed synchronization of operations, all of the tool changer commands are initiated from the part program in the form of m-codes. (Due to of the vintage of the cnc system, it was easier to add the necessary control program for the tool changer to the operatiang system of the shop-floor processor than it was to alter the executive program of the CNC computer.) Coordination of the various computer controlled cycles is accomplished through the use of a series of machine stop commands in the part program and appropriate continue commands from the shop-floor computer. This "handshaking" process assures that one operation is completed before another is initiated. (The same functions could be achieved via a hardware interface, but since the shop-floor computer was already available it was easier to perform the task with software.)

During execution of the tool change cycle, operations are controlled by the occurrence of the appropriate m-code in the part program, which is followed by the previously mentioned machine stop command. After a particular tool change m-code command is read from the part program and output to the shop-floor processor, the CNC system halts its operations and waits for a subsequent cycle start command. This command to continue part program execution is generated by the shop-floor computer when it receives a signal from the tool change computer that the operation defined by the initiating m-code has been completed. At this point, operations are resumed under the specific control of the part program. (If this system were designed with a modern CNC system the tool change computer would be superfluous because the operation of the machine control unit could be readily altered to perform this task.)

An advantage to this type of system is that it is relatively simple as well as being easy to implement. The only communications that must be handled by any given computer are one level up or down in the heirarchy, there are no horizontal lines of communication. A disadvantage is that there can be no sharing of resources so that if one processor is unavailable then all operations at its level and below are halted. For the above example, the shop-floor processor acts as a host to the CNC system and the tool changer computer. If the shop-floor computer is not accessible from these other two CPUs then manual intervention is required to continue the machining operations.

Network Communications

In general, there are three basic categories of computer communication operations that can occur. Centralized data communications refers to the transfer of information within a particular computer system. One example is the data that is transferred from an input terminal to the main data processing computer (host) and on to a disk memory or output printer. Figure 8.4 shows the equipment interconnections used with a centralized control system for a machining cell. In this instance, the cell control system consists of a single computer that is in charge of everything. This host computer manages a data base and implements all of the cell control functions, including supporting the data requirements of the individual machine control systems (the application) and executing the necessary logic/rules for the communication task (protocol).

In contrast to the centralized communications approach, the decentralized communications system involves information transfer between computer systems in a master-slave relationship. While the host computer is still overseeing everything, there is a broadening of responsibilities at the lower levels of the network. Figure 8.5 depicts a networked cell control system in which the cell controller is responsible for the cell control functions, the data base management and the machine application software. However, the protocol section of the machine-dependent software is contained within local network interface units (NIUs). When the cell controller wishes to send a command to a particular machine it sends

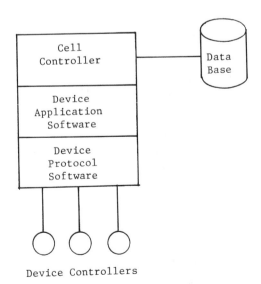

Device Controllers

Figure 8.4 Centralized cell control system.

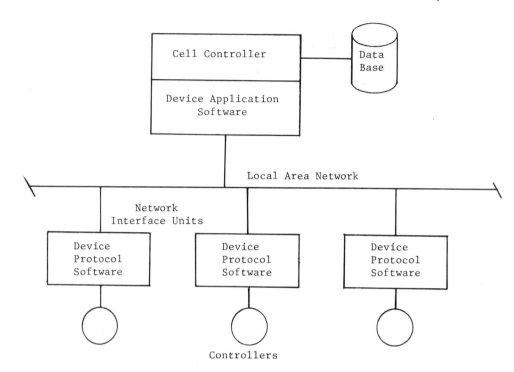

Figure 8.5   Example of a network cell control system.

a message over the local area network to the NIUs which route the message to the appropriate machine.

Distributed communication systems involve network communications between self-contained, autonomous entities. In this situation, there is no master-slave relationship and the computer systems operate in a cooperative fashion. In actual practice, a communications network frequently involves all three of these types of information transfer operations at different levels of the system. Therefore, it is often ambiguous to attempt to classify a system as belonging to only one specific category of operation. For example, a group of host processors may be interconnected through a distributed communications network, while each of them is also serving a variety of subsystems. These subsystems may involve both centralized and decentralized communications at different locations within their heirarchy. Even the lowest level of a heirarchy, such as a machine tool control system, may incorporate multiple CPUs that work together to provide the necessary data processing capability (servo control, command interpolation, I/O, etc.).

In the previous section, an example was presented that described the information transmission requirements associated with a single machine tool. In this case, the communication processes occurred through the use of a relatively simple computer network that supported the needs of the DNC machining operations. In contrast, Figure 8.6 shows a more complex distributed control system that is currently in use at a light truck manufacturing factory [3]. In this instance, there are special applications processors located on the LAN that gather data from the other nodes and perform supervisory control tasks or provide information directly to an operator. An advantage to this system design is that other functions may be added to the control system through the addition of new processors without causing significant disruptions to the existing functions.

In the arrangement depicted in Figure 8.6, a variety of processing nodes are attached to the central network in a bus architecture. The storage, routing and alarm nodes are processors that maintain the static data base for the complete system. On request from the various nodes, these processors down load the appropriate computer code operating instructions and other data base information. In addition, these nodes also monitor themselves and the other nodes in the system and notify the system operators if problems occur. The equipment interface nodes are responsible for communicating with all of the plant-floor equipment.

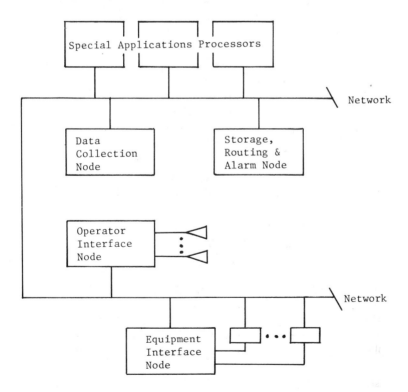

Figure 8.6 Computer network used in light truck manufacturing facility (From Ref. 3).

These nodes store the dynamic information associated with the plant-floor equipment and compare the monitored values of the various parameters with alarm limits so that alarms and warning messages may be generated whenever appropriate. The operator interface nodes provide a color graphic representation of the equipment and process status. These nodes provide a zoom linkage between the graphics pages so that the operator may obtain detailed information about a particular subsystem without a significant time delay. The data collection nodes collect dynamic data from the equipment interface nodes, perform data reprocessing and store the data on local disk memory for use by the other nodes.

Figure 8.7 shows another control architecture [3] that is designed to support the Automated Manufacturing Research Facility (AMRF) at the NBS in Gaithersburg, Maryland. The AMRF is a research facility that was established to facilitate the development of a small batch manufacturing system that would support study and experimentation in automated metrology and interface standards for the computer integrated factory of the future. This integration of computer systems with machining operations is intended to allow enhanced system control as well as flexibility in system configuration and the manufacturing of workpieces. The integration of the various manufacturing process is designed to occur through heirarchical task decomposition and real-time sensory-interactive concepts developed at the NBS [4]. In this approach, the system control modules are arranged in a heirachy in which each controller takes commands from only one higher level system, but it may direct the activities of several systems at the next lower level as shown earlier in Figure 8.7. The long range goals of the manufacturing facility are input into the system at the highest level of the hierarchy and are decomposed into subgoals that are executed at that level or lower levels. Status information, based on real-time sensory data collection, is obtained at each level and passed up the ladder as feedback to the next higher level.

The transmission of information between the various computing processes and systems within the AMRF takes place via a communications network that is transparent to the individual computing processes. Each data processing operation writes to and reads from memory locations that are universally accesssible by all the different processes (common memory locations). Each individual process that has information that may be needed by another process stores this data in a "mailbox" location in the database. Thereafter, this information may be obtained by any process that needs the data to complete a task. A network interface process provides the communications protocols, or rules, that enable these data transmissions to occur without degradation of the information content.

For simplicity, the remainder of this section will focus on the AMRF cell control system (CCS) used with the turning, horizontal machining and material handling workstations. (However, the principles are similar at each control level of the manufacturing system hierarchy.) The function of the CCS is to manage and coordinate the performance of all part production and support tasks at the workstations that fall within its area of control. In performing this task, the CCS is

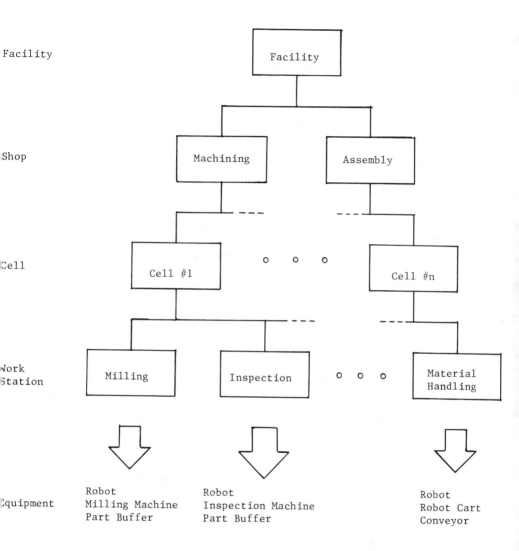

Figure 8.7 Control architecture hierarchy for AMRF (From Ref.4).

required to produce parts and to implement changes in the configuration of the manufacturing system as commanded by the next level up in the control heirarchy.

The system handles requirements for manufacturing particular workpieces as work requests that are organized as entries in work queues. This listing of future job requirements is maintained in a data base. To manage the production of workpieces, the CCS removes entries from these queues, schedules the required tasks at the various workstations, monitors the execution of these tasks and provides continual feedback reports on the operational and task completion status of each work station to the CCS host processor. Also, the CCS coordinates the restocking of raw material and the pickup of finished products at each work station, and the transfer of in-process inventory between the necessary work stations on an as needed basis.

The CCS controls the individual work stations through the use of three software modules: the queue configuration manager (QCM), the scheduler and the dispatcher. Inputs to the CCS consist of commands from the next higher level control module and feedback from the next lower level module. Outputs from the CCS are made up of the feedback to the next higher level and commands to the next lowest level. All data shared between cell controllers, other than commands and feedback, is stored in and exchanged via common data bases.

The highest level function of the CCS is the execution of the commands that come from the CCS host. These commands require that the CCS manage and coordinate the activities of those workstations that are under its control. To accomplish this, the CCS utilizes the QMC module to provide the interface between the CCSs host and the CCSs internal scheduler module. When a new order is ready for processing, the QCM creates an entry in the cell job queue, determines the tasks necessary to meet the request, assigns those tasks to the appropriate schedulers and reports that the job has been accepted by the system.

The scheduler software modules interface to the QMC module above and the dispatcher modules below. The task of the scheduler system includes selecting the next task to be processed, initiating the processing of that task via the appropriate command to the corresponding dispatcher module, clearing the scheduler queue of all completed or cancelled tasks, and updating the status of any new or in-process tasks.

The function of the dispatcher module is to dispatch the selected task to the appropriate workstation and to monitor the system's progress toward the completion of the assigned task. This hierarchical arrangement of control modules has been demonstrated to be very useful in defining the interfaces between the different processes. In addition, this approach allows the modular components to be developed independently of each other and then integrated into a complete system.

Based on the results demonstrated by the NBS in the AMRF, it appears that a system of interface standards may be possible for interconnecting computer-aided design, process and production planning, scheduling, materials transport, and control systems with machine tools, robots, sensors and sensory processing, data bases, modeling and communications systems. An advantage of having these standards is that it may become practical to implement computer-integrated manufacturing systems in an incremental fashion using equipment from a variety of vendors. Incremental automation is particularly attractive because it is often much easier to justify and pay for incremental changes in system configuration than it is to redesign an entire operation. However, to achieve success in this incremental fashion requires that the basic system be well suited to the necessary task and that the interface points required for the mechanical and electrical modifications are readily available.

ISO Reference Model

The goal of a data communications system is to provide a cost-effective means for permitting data processing programs to communicate, through the use of equipment such as the telephone system or other networks, when the physical data processing equipment is distributed over a wide range of geographical locations. Since networks are used to interconnect this variety of computing systems, it is not unusual for different types or brands of computers to need to share information and data processing resources. This type of a distributed data processing arrangement requires operating standards, or rules, if the communication process is to be effective. The ISO reference model is an attempt to provide a level of standardization to this task. Progress toward defining this model is a result of ISO Technical Committee 97 (data processing) which created the Study Group 16 (Open Systems Interconnection) to examine the problem of standardization of distributed systems. In order to prepare the United States position, the American National Standards Institute (ANSI) created Project 300 to deal with distributed systems. The lower levels of the model deal with the physical communication medium, while the higher layers involve the logical connections between two communicating computer programs. At present, ISO has established standards for the lower layers of the reference model and work is underway on the other layers; however, it is uncertain when this effort will be complete.

Another reason that a data communications system model is desirable is that its existence makes it easier to discuss the different facets of a distributed system. In addition, it facilitates the comparisons of the various features provided by different system architectures. Typically, a model consists of a series of functional layers, or "black boxes," that solve specific tasks. The results of the actions undertaken within these layers is then communicated between the layers so that the overall data processing objectives are achieved.

In the ISO reference model, the lowest level is the physical layer. This layer provides a communications path over a physical media for open systems interconnection between different data processing systems. It coordinates the physical connections between the various data link entities and has specific mechanical and electrical characteristics as well as procedures for controlling the data bit transmissions. This layer serves as an interface to the next layer up in the heirarchy, the data link layer.

The data link layer is located between the physical and network layers, and serves as an interface between these two entities. The purpose of the data link layer is to provide the functional and procedural methods to enable reliable transfer of data blocks across potentially unreliable communications media. The objective of the data link layer is to transmit data blocks over this questionable channel and to detect and correct the bit errors. Techniques involving redundant data transmission or retransmission of information are utilized to control the potential errors. However, this is done at the expense of dedicating some of the communications channel capacity to this error detection/correction activity which leaves a reduced capacity for message transmission.

The network layer of the ISO model provides communication services upward to the transport layer and downward to the data link layer. In effect, it makes the lower level physical channels and network topology links transparent to the transport layer so that the specific protocol implementations required at the lower layers do not need to be considered at the higher layers. The network layer is responsible for the routing and relay functions over the switched communications channels. These routing functions may be based on predetermined, unchanging routes or they may be based on the changing conditions within the network.

The transport layer is the principal interface between the user layers and the network layer. The lowest three layers of the model provide data transmission and routing functions. The transport layer assures reliable and efficient end-to-end information transport service between users. The transport protocol provides a set of commands for user to user communications as well as handling message sequencing and error detection when appropriate. In addition, the transport layer assembles network messages into a size that is most efficient for network transport and deals with information flow control and network conjestion.

The session layer, directly above the transport layer, is the first of the user layers. It provides user oriented services to the presentation layer, above, and the transport layer, below. (In some network architectures the session layer is not well defined, and the session layer management functions are performed as a part of the application layer.) At the session level, one program is able to initiate communication activities with another program, on another machine, by requesting a session with the other program. If the request is successful, the other program is ready to interact with the first one, and messages are sent and received through the transmission across the various physical links. Another important function of the session layer is to make the transport connection failures transparent to the higher layers.

The presentation layer lies between the session and application layers and is only concerned with the form, format and syntax of the information. It is responsible for code and character set conversions, modification of data formats, and the coordination of source, destination and transmission syntax. For instance, a program on one computer may be written in FORTRAN while another program on another machine may be written in COBOL. Since these two languages utilize different data formats it is necessary to perform the necessary data conversion to permit the desired communication to occur in a meaningful fashion. This layer is concerned with form of the data but not with the meaning of the information.

The application layer provides an environment for the operation of network and distributed operating systems as well as distributed data base and data processing systems. Typical application layer implementations include remote resource sharing, electronic message (mail) systems, file transfer systems, and distributed data bases. This is the point at which the information content of the messages becomes important as opposed to the need to perform an accurate data transmission that exits with the lower levels.

Manufacturing Automation Protocol

In a manufacturing facility that has progressed through a continuously changing evolutionary process of automation, it is typical to encounter a variety of machine controls, programmable instruments and computer work stations of various vintages that are manufactured by different vendors. When attempting to combine these entities into a computer-integrated manufacturing system it is necessary to install special hardware and software to assure compatibility before network communications can even be attempted. One alternative to this approach is to select a single vendor as a source for all equipment. However, assuming one vendor can be located to satisfy all of a facilities requirements, it is not necessarily desirable to be locked into a specific product line. In order to avoid having to follow this approach as a means of assuring equipment and software compatibility, a standard for factory floor communications is needed.

In 1980, General Motors (GM) created a task force to identify communications standards that would permit data communications between different vendors' equipment. A primary objective of the project was to evaluate the international Open Systems Interconnection (OSI) reference model for use in factory automation. (Open systems are ones that are open to each other through their mutual recognition and use of applicable standards. However, these systems may not be mutually compatible without a standard communication interface.) The OSI model describes the functions that each network device must be capable of performing, in a manner that is independent of a particular system design. However, because the OSI model only involves functions and not protocols, compliance with the model does not mean that communications can occur between equipment that is manufactured by different vendors. In 1981, the GM task force proposed the use of the OSI model in the implementation of a local area network that would allow direct communication between different data stations. GM then lead the way in the development of the Manufacturing Automation Protocol (MAP). Furthermore, GM stated that it would purchase only equipment that was MAP-compatible. At present, GM is utilizing MAP version 2.1 in their production of automobiles and version 3.0 is ready for demonstration.

Manufacturing Automation Protocol is a hierarchical system that utilizes a network of coaxial cables to connect computer-controlled equipment on a factory floor. The individual stations are usually connected in either a bus, ring or star arrangement as shown in Figure 8.8. For a bus configuration, the stations are linked through a single transmission medium in a pattern like the branches of a tree. Any station may transmit or receive information without the intervention of intermediate network nodes as long as the data transmission link is intact. In the ring network, the stations are wired together in a continuous ring configuration. As the network messages move around the ring, the different stations check the address that is included in the message to determine the intended recepient. Only those messages with a particular station's address are retained, the others are passed on to the next station. In a star network, the individual stations are wired to a central hub that routes the messages to the proper location.

Manufacturing Automation Protocol is a standard that addresses the problem of communication among diverse, normally incompatible devices. The MAP specification describes a communication approach that consists of a seven-layer heirarchy that is based on the ISO model. It is both a philosophy of device integration and the specifics of required communications protocols. Unfortunately, much of the MAP literature deals with a detailed description of the system's lower layers (which is not of much interest to many users) as opposed to information on how to apply MAP. In addition, MAP networks are neither the only solution nor the least expensive approach to dealing with the problem of factory communications. Proprietary networks are available that may be more applicable in areas where standards do not exist or when the standards that do exist do not encompass the necessary communication goals. Also, it can be quite expensive to switch from an existing network to MAP. At times, the use of a private network may be more suitable in locations such as a manufacturing cell. If desired, this network can then be interfaced with a MAP backbone that permits communications to other processes or sections of a factory. The devices that provide this link between the private network and the MAP network are called gateways.

One GM MAP installation is a prototype manufacturing facility that produces axle components for front wheel drive vehicles [5]. The control system hierarchy at this production plant consists of a factory scheduling system, a factory control system, cell controllers and a material handling controller. The factory scheduling system develops the production schedule based on order demand, factory inventory, cell status and expected receipts. The factory control system keeps track of all the resources and translates the production schedule into the specific

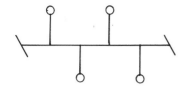

(a)   Bus Network

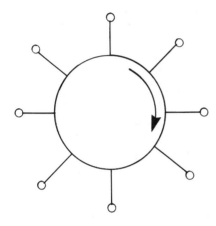

(b)   Ring Network

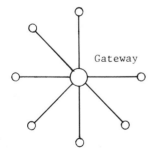

Gateway

(c)   Star Network

Figure 8.8   Typical network architectures.

activities that are required to support this schedule. This system gives instructions to the cell controllers and the material handling controller while receiving status and monitoring data from these same systems. The cell controllers coordinate the activities at each individual cell while the material handling controller directs the activities required to carry out the material handling commands. MAP is used for all communications between these functions as well as the various components of the factory control system which involves more than 40 distinct computer systems.

Figure 8.9 shows a sketch of another Map network that is being used in the automotive industry as part of a retrofit program to increase productivity and quality [6]. This sketch shows the straightforward connection of the host computer support system to the MAP broadband network. Commercially available interface adaptors are used to connect the robot controllers to the MAP network. These interface adaptors are printed circuit boards that are installed in the robot control cabinet and are transparent to the systems that they support.

While MAP can offer significant advantages to the user, it is still not a mature specification and potential users should carefully consider all the objectives that a MAP installation will be required to meet [7]. Hardware devices needed for a particular application may not support MAP and other problems may arise due to an incompatability with specific software packages. For instance, MAP may be a good choice for communication between physical devices but inappropriate

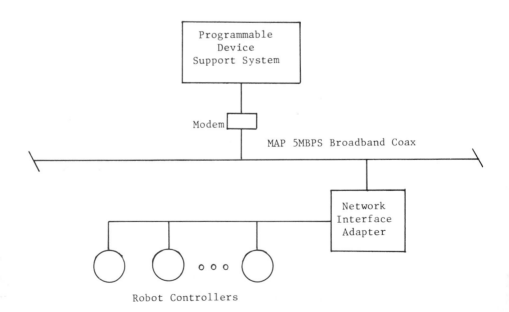

Figure 8.9 MAP network used in automotive industry.

for use between cell controls and higher level factory controls (or vice versa). MAP should not be forced into a software architecture when it is not appropriate. If it cannot be utilized as the backbone of the communications scheme then another approach should be employed. At the same time, if it is the backbone of a network, then it should be used whenever possible to avoid communication protocols across system boundaries.

Another point to remember is that MAP is not failure-proof. In spite of the extensive error checking protocols that are embedded within its layers, MAP is still vulnerable to the problem that exists with all communication systems. When a host computer crashes or an application process fails, messages can be totally lost if they are in the pipeline between the sending and receiving processes. In a monitoring application, the risk of an occasional lost message may be acceptable, but this is usually not the case in a controls situation. In this instance a protocol must be placed on top of MAP to detect and resend the lost information. One approach for accomplishing this is to uniquely number the messages and employ a handshaking protocol that detects the existence of missing data.

System Reliability Enhancement

The use of computers in manufacturing operations continues to spread because of the benefits that they provide in data manipulation. However, with the increased dependence on these computing machines also comes increased vulnerability to having operations disrupted due to computer system malfunctions. It must be recognized that any computer system will eventually fail and the full impact of this failure must be understood. In order to avoid having a bottleneck such as this disrupt an entire operation, provisions must be included in the system to permit manufacturing operations to continue while some portions of the network are out of service for maintenance. At various levels of the network hierarchy this may be accomplished in different ways. At the upper levels of the network, duplication of data pathways and available processors provides the needed insurance that a contingency can be accommodated with little or no penalty in system performance. At the lower levels of the network, such as within a machining cell, there may be only one host that is connected to the individual machines. In this case, an alternate data path may not exist and it is necessary that the machine controls have a degree of self sufficiency that allows them to operate for extended periods without interaction with the host system. This may be achieved by providing sufficient storage capacity within the individual machine control systems to accommodate the needed part programs plus an alternate method of loading additional part programs via a portable solid-state memory unit or through a peripheral that utilizes a backup tape or disc memory.

The reliability of networks should be viewed in two aspects. One viewpoint is the physical integrity of the data transmission while the other is the security of the data. Data may be corrupted through a failure in hardware, software or an operational procedure. Also, the possibility exists that the information may be misinterpreted during processing. Data integrity can be inhanced by providing multiple paths over which information may flow to its destination. Another approach is to utilize a combination of serial and parallel data paths so that most crucial data links are protected by a backup link. In addition, the use of error detection and correction schemes is a means of accommodating intermittent system failures that result in an occasional error as opposed to the loss of a complete portion of a network. Network security is important to protect the system's information from being corrupted as well as to safeguard proprietary or limited access data. Industrial espionage exists in many industries while banking institutions are potentially vulnerable to criminal actions. Physical security is achieved by limiting the physical access to data processing areas and information resources. In addition, user identification schemes exist which range from a simple password to more exotic techniques such as fingerprint or lip print readers or dynamic signature analysis. Also various encryption techniques are available to scramble data so that even if it is intercepted it is difficult to understand. Data privacy is another important consideration. Unfortunately, designing systems that provide adequate protection to personal data is very difficult especially since a disgruntled employee can do so much damage in a data processing environment. However, this does not lessen the need for safeguards. (Perhaps the best way to deal with this problem is to avoid having personal data in a data base system whenever possible.) A good security system for a computer system as well as a computer network involves the use of an audit trail. This approach utilizes a data file that keeps track of any significant event such as when users log on or off of a system as well as when sensitive areas or transactions are successfully or unsuccessfully invoked.

References

1. J. Lincoln, CAD/CAM supplying the competitive edge, *Manufacturing engineering, 97*:6 (1986).
2. R. A. Piety, Intrabuilding data transmission using power-line wiring, *Hewlett-Packard Journal, 38*:5, pp. 35–39 (1987).
3. J. A. Vrba, CAM for the 80's—Distributed Systems Using Local Area Networks, Proceedings of AUTOFACT 6, Anaheim, California, pp. 10–14–10–27 (1984).
4. J. S. Albus et. al., A Control System for an Automated Manufacturing Research Facility, Proceedings of Robots 8 Conference and Exposition, Detroit, Michigan, (1984).
5. D. Richardson, Implementing MAP for factory control, *Manufacturing Engineering, 100*:1, January (1988).
6. K. G. Hughes, Implementing MAP on the factory floor, *Control Engineering, 33*:10, pp. 156–158 October (1986).
7. D. Richardson, Implementing MAP for factory control, *Manufacturing Engineering, 100*:1, January (1988).

CHAPTER 9

System Software

Introduction

Computer systems consist of a set of physical entities that perform tasks through the transmission of electrical signals and the instructions that organize this electrical activity. The physical devices are referred to as hardware because of their relative permanence, while the instructions, or operating programs, are called software because of their temporary nature. This software may be characterized into two general categories. One type of software is the set of instructions that make the computer "work" properly, or act like a computer, while the other type of software deals with the specific application with which the computer system is currently involved. The first category performs tasks such as transferring information files from one memory location to another and is called the operating system software. The other category deals with a particular application such as monitoring a process characteristic and controlling a process adjustment based on the result of a calculation that involves the monitored attribute. This type of software is called applications software. Without the operating system software a computer is just a "hunk of iron," it cannot even understand the instructions programmed within the applications program. Without the applications software the computer may have tremendous potential power but is not being applied in a useful fashion.

Regardless of the type of operation involved, the computer software, or code, is an often neglected portion of a manufacturing control system at the time when initial planning activities are underway. When this oversight occurs, it is only after things are somewhat finalized from a mechanical standpoint that the software issue gets sufficient attention. Since this code is the electronic grease that makes the automated factory run smoothly, there are tremendous opportunities available by giving software sufficient attention in the early planning stages as well as throughout the project. In contrast, there are also numerous pitfalls to be encountered if the overall software problem is not given sufficient attention. These potential opportunities that may be pursued are limited only by the available computer hardware and the imagination of the system designers. Also, in many instances the computer hardware has evolved far faster than the software needed for particular applications. In the typical manufacturing environment, the computer hardware is rarely the limiting factor; however, the limited imagination factor can be very significant for reasons such as inexperience, conservatism, distrust, and so on.

In new applications involving currently unavailable or untested computer software, it is typical to significantly underestimate the amount of manpower resources that will be required to accomplish the generation of a viable computer code for the desired system. This pitfall is frequently encountered because there are relatively few individuals in industry that have both the in-depth understanding of manufacturing processes and the necessary background in computer applications required to accurately estimate and manage the task of marrying the software to the application. Also, even though there may be very capable individuals involved in the tasks required to generate the software and to run the manufacturing facility, there still may be communication difficulties that can lead to very significant problems. To avoid these obstacles, a rigorous definition of the overall task must be agreed upon before programming activities begin. In addition, subtasks must be defined, when appropriate, and the linkages between these tasks clearly established so that the programming can be accomplished by multiple individuals without confusion over how the pieces of the system interact. This software modularity is also a desirable feature for dealing with future system modifications. Inhancement or expansion of a system through the incorporation of additional machines or work stations can sometimes be achieved by modifying data in system definition tables. However, in other instances it is necessary to employ more extensive modifications to the system programming. These expansion efforts are most easily accomplished if an initial system design approach was utilized that is based on task modularity.

Other crucial requirements for control of the software "beast" are accurate, detailed documentation and upward compatibility of the code within a supplier's line of computer products. Software is a resource in terms of both a purchase price or other acquisition cost and as a capability that exits within a particular computer system. As such a resource, it should be treated appropriately in relation to the value that it represents to an organization. Conventional products are manufactured to a particular set of tolerances that define the quality of a unit; however, this quality assurance approach is not always followed in the management of computer software. Obviously, the frustrations involved in attempting to maintain or upgrade undocumented software are only surpassed by encountering erroneously documented software that does not perform the task that is required for a current application. In many cases, it is better to start from scratch and completely reprogram the needed computer code because of the problems that are inherent in trying to deal with undocumented or incorrectly documented software. Unfortunately, that option may not always be available and in these instances the expenses associated with fixing things can be a significant threat to the successful completion of a project. Other difficulties that may be encountered, are that the existing software may not have been upgraded as hardware improvements were installed or else patches were employed to provide a temporary fix to a current problem. In this case, the unfortunate result is that new computer hardware is used for the execution of old applications programs that are not structured to take advantage of the improved features that are available with the current computer system.

A related problem is encountered when the existing computer hardware is exchanged for a different system that is manufactured by another vendor or sometimes even when obtaining another system that is supplied by the same vendor. Since a significant amount of resources is usually invested in existing application programs, it is not unreasonable to be interested in transferring these available programs to the new computer system. (In addition to the expense of preparing and testing the new code, there is also a trust factor that is involved on the part of the computer user.) Previously, this transportation of application programs was very difficult to accomplish because the code that performed special purpose tasks with one computing machine and its particular operating system would not work with a computer package supplied by another vendor. The systems were "closed" to each other in that mutual communications were difficult if not impossible to achieve. There was often a language gap as well as other problems that inhibited the user from changing systems. While this may have some limited benefits in assuring customer loyalty, it is actually a detrimental situation for everyone when a broader point of view is taken.

The concept of an open architecture is in direct contrast to the previously discussed closed situation. This approach is designed to avoid many of the compatibility problems inherent with intermixing computing equipment that is provided by different suppliers. An open architecture is one in which the computer system readily permits the integration of user-written programs and this application code is easily transferred (portable) between systems provided by different system vendors. One example of this type of architecture is the UNIX operating system developed by AT&T. UNIX, and other open systems, can easily accommodate modularity so that building blocks can be arranged in whatever manner the programmer desires.

Another unfortunate but ubiquitous fact-of-life in the computer software world is the continual flow of "it sure would be convenient if we could only . . . "

ideas that inhibit the finalization of the system configuration. Some changes in system functions may be relatively easily accommodated after the initial system definition has been specified. However, other desirable modifications must be delayed for later versions of the system software or the project will drag on indefinitely without completion. In many instances, it is not unusual to encounter someone who assumes that their proposed changes in system operation should only involve the modification of a few computer instructions. ("After all, it's only software, right?") Unfortunately, it is not always this simple and an in-depth analysis may be required to determine what effect the change in one portion of a system has on other parts of an operation. Unpredicted effects of innocent code changes have a way of becoming very significant when they are least expected and when they are particularly inconvenient.

This chapter deals with the computer software associated with numerical control systems (including the software used to direct the individual control units), monitoring of manufacturing systems, data manipulation and process modeling. The operating system's software that is mentioned briefly is not covered in detail here because it falls within the area of how a computer works. That subject is considered to be outside the scope of this test. However, the executive program software used with a computer control unit, such as a CNC system, is roughly equivalent, on a smaller scale, to the operating system software that runs on a larger computer.

Computer Control Executive Programs

Manufacturing facilities that utilize one or more computers to control the operations of various in-house systems are dependent on both the computer hardware and the software to achieve the desired form and level of system control. (Without the correct software the computer system is just a set of physical components that is more of a detriment to production than a benefit. In addition, one of the most irritating things about computers is that they are so good at doing their assigned tasks when everything is working correctly so that it is very frustrating when there are system malfunctions—it is relatively easy to become dependent on these machines.) In the case of dedicated computer control systems, the system software or executive program may be difficult to separate from what was previously called applications software. This composite computer code is tailored for a specific application such as machine tool control, parts handling, assembly line control and so on.

For a relatively simple task, the applications software may be the only software that is executed by the computer control unit. The assigned task may only be to continuously examine the status of a few input signals, compare them with target values, perform a few mathematical operations and output a signal that is descriptive of the state of the process. One example of this type of system would be a computer that had the sole function of closing the position loop for a machine servo system. In this situation, the elapsed time that occurs between the output updating operations is only a function of the time required to execute the computer code because the computer is essentially dedicated to a single task, (this may be a requirement for control system stability in some applications). In a more complex system, the control software may be required to execute multiple background tasks while communicating with a number of other computers (through the use of operating system software) to provide the information needed for the data processing requirements on the other machines.

An example of a somewhat complicated computer control system is the CNC unit used on a multi-axes machine tool where individual processors are utilized to control servo systems, I/O status information, communications, and so on. In this instance, the resources of the central computer are shared by a number of applications, in a timely manner, so that each task is serviced without causing excessive delays. This may be done either by giving each task an equal opportunity to utilize the computer's resources or it may be done by setting up a priority ranking system where some tasks are judged to be more important than others and are assured of receiving attention within a specified time period. Examples of tasks that might be judged to deserve high priority include closed-loop control algorithms and error detection. The maximum time period that would be allowed between updating these tasks might be limited to 10 milliseconds or less. An example of a lower priority task would be the output of the display information for the status indicators associated with a manufacturing process (run, halt, manual, automatic, etc.). A one-second maximum cycle time for information updates would not cause problems in the control of the system's operation for many applications, especially when the delay is only in illuminating a status light and does not delay any control actions in response to the system status.

Today, the control systems used in the machine tool world have evolved to the point that the executive program, or operating system software, handles all of the tasks associated with the control of a particular machine. With the early NC systems, these functions were executed by hard-wired logic elements that were made up of devices called switching circuits. In these systems, the logic circuits were constructed from vacuum tubes and discrete components such as resistors and capacitors. Because of the size of the individual circuit elements, these NC systems were rather large and contained a lot of individual components so that the maintenance requirements were significant. The subsequent development of the transistor and integrated circuits simplified things because a larger number of logic elements could be packaged onto a single electrical component. This provided a reduction in size requirements as well as reduced maintenance needs even though the capabilities of the control system were significantly expanded.

With the substitution of the computer circuitry and software commands for the hardwired logic boards, the NC system became a CNC system. Where the previous control logic algorithms were determined by the hard-wired connections between the logic elements within the NC unit, now a set of computer instructions is utilized to achieve this purpose. In addition, the internal complexity of a single integrated circuit is greater than what was initially contained within the entire early NC unit, and today's computer control systems contain a large number of these devices which are used to execute the instructions in the computer and to inferface the computer to the manufacturing process.

When the early CNC systems were introduced, they quickly took over the machine tool control market because they offered a significant price reduction in comparison to the existing NC systems. However, the CNC systems did not allow the user much addition flexibility unless the executive program was returned to the vendor for modification. In addition, not many vendors were very interested in doing specials, because it was simplier to maintain a standard product line. Because of the inflexibility of the older NC units that the market was used to, this limitation was not recognized as being a problem. However, as the users matured, it gradually became obvious that being able to tailor a control unit to a specific application had significant advantages and that having to obtain a new executive program from the vendor was not an efficient way to proceed.

Once users discovered the advantages associated with having the ability to easily alter a control system's configuration, they also wanted a more appropriate means of implementing system modifications. Later CNCs have largely eliminated this stumbling block and permit the user to prepare special software or subroutines that are executed by the CNC along with the basic operating system software. These subroutines are written in a language that includes if-then and go-to type commands. In addition, the capability exists to modify a system's I/O definition as well as to perform tasks such as the substitution of a variable parametric expression for the numerical value of any part program word, implement canned cycles or utlize higher level math functions (square roots, trigonometric functions etc.).

On a CNC system, the equivalent to the applications programs that were mentioned earlier is a set of machine commands called a part program. This set of instructions is interpreted by the CNC system executive program and translated into the required electrical signals to cause the machine tool to perform the necessary motions or actions that are needed to fabricated the desired product. This part program may be thought of as a higher level language since it contains machine commands instead of computer instructions. However, the interactions required of a machine's axes may be quite complex to produce many products so that it is often difficult to prepare the part program using manual techniques. Instead, the part program is typically generated through the use of a high level software package called a processor which converts the geometric definition of the desired part into a data set that relates to the machine motion segments (cutter location or CL data) used to manufacture the workpiece. However, this data set only solves the geometry portion of the problem. It does not account for the differences in particular machines such as the position resolution of the individual machine axes, any optional special features or limitations on feedrates and spindle speeds, and so on. In order to obtain a usable part program, another translation program called a postprocessor is used to convert the source code that is output from the processor into machine-readable commands that are tailored to a specific machine and control system combination. One disadvantage to this approach is that many postprocessor programs may be required for a large machine shop since they tend to be machine-specific. Maintaining a large number of postprocessor programs can be frustrating and expensive. In addition, a significant amount of mainframe computer resources may be required to perform the postprocessing tasks.

One approach toward alleviating the post processor problem is addressed by the EIA standard RS-494. The technique involved is to use a translator program that transforms the CL source code into a standard format. This translated code is then executable by any CNC with an internal postprocessor that can execute RS-494 compatible code. This approach eliminates the need for many different postprocessors at the expense of placing an additional data processing burden on the CNC system. This type of control system has been used successfully in the aerospace industry but, at present, not many control system vendors appear to be interested in pursing this technology.

Another technique used to avoid the necessity for a postprocessor is to create a software package that operates on the machine tool control system which allows the user to define the workpiece using common geometric terms. In effect, the user friendly approach permits the operator to input geometric terms and machining parameters that the CNC system translates into a usable part program. This is quite convenient for relatively simple jobs, but it is inadequate for more complex tasks. For example, this programming method would be suitable for use with a two-axis milling machine that was utilized for general shop work, but it could not support multi-axes contour milling applications.

Error compensation is another aspect of machine tool control that is supported by the control unit executive program. Early CNCs offered standard software features that permitted corrections for characteristics such as cutting tool offset errors, leadscrew positioning errors, fixture location errors and cutter size errors. In addition, special purpose software was available from the control system vendors to accommodate cutter shape errors. However, these functions have gained wider usage with the advent of the CNC units. Also, current generation CNC systems permit additional options. These added features include the automatic use of on-machine gaging for the inspection of a workpiece, the automatic adjustment of the machine offsets to accommodate a size error that is detected prior to the last machining pass or on a previous workpiece, and communications protocols that support DNC.

Direct Numerical Control Software

The software that is required for direct numerical control (DNC) operations is located primarily on a host computer and is able to provide DNC support to a number of independent work stations. This software operates on the host computer because this is the focal point for each of the individual numerical control systems. If the machine tool control system is capable of handling the necessary bidirectional communications with the host then no additional interface equipment will be required. Otherwise, the host must communicate directly with an additional shop floor processor that passes information back and forth from the machine tool control through the control system's serial interfaces (which are usually used for the tape reader and tape punch). In either situation, additional interfaces also may be required to provide automatic remote control of the control system's functions. The hardware and software must be available to allow the shop floor processor to push the buttons on the control system as it simulates the actions of the manual operation, or else a machine operator must be available to initiate and supervise the activities of the system. Manual operations may be quite acceptable if the major DNC activities consist of occasionally down loading part programs and up loading status information. However, if an operator is not always present or if portions of the system's operations such as on-machine gaging and adjustments require automation then the interfaces become more complex.

In a typical installation, the DNC software provides the communications interface needed to permit the transfer of part programs and other data between the host and local computers, enables remote control of the NC machines and provides status monitoring. In addition, in some situations the machine tool control system may be treated as a virtual terminal for the purpose of communications with the host computer system. This arrangement means that the control system operator can use the control system keyboard to log onto the host and to do whatever operations might normally be performed with a conventional terminal (assuming the appropriate keys are available on the control system's keyboard). Also, directory manipulation commands and machine offset update activities are often permitted in conjunction with part-program execution operations as long as these modifications do not impact data records that are currently in use.

If the DNC system is purchased as a complete package then the user-defined features will be set up as defined in the purchase specification. Otherwise, it is necessary for the user to configure the system to meet the particular needs that are being addressed. In either event, it is desirable that the user have ready access to the files and system tables that define the system's mode of operations. This is because it is very likely that at some future time it will be necessary to modify some aspect of the system.   .

Process Monitoring Software

As described in an earlier chapter, process monitoring activities can provide important feedback information that may be used in establishing the quality of a manufacturing operation. This process of monitoring the quality of an operation while products are being manufactured is achieved through the characterization of appropriate system parameters. In order to accomplish this objective, a suitable combination of measurement transducers and data analysis techniques are required. The intent of this procedure is to infer the quality of the manufacturing process through a statistical interpretation of the output from the measurement transducers. (It is assumed that the monitored information has been previously proven to be indicative of the condition of the key process parameters.)

Various measurements transducers have been described in a previous chapter and for simple manufacturing operations it may be possible to monitor these devices using manual methods. However, for more complex monitoring needs, the use of computers is frequently required to pre-process the measurement signal, collect the information in an appropriate fashion and to analyze the resulting data. While this more complicated task could be accomplished using manual methods, the use of the computer permits the completion of the data processing in a timely fashion. This allows the resulting information to be available while the manufactured products are still at a point in the fabrication cycle at which process adjustments can be performed that will result in improved quality. Significant delays in information collection and analysis result in data that describes the condition of the operation for a given batch. While this is valuable information from a historical standpoint, it is of relatively little value for improving the level of real-time quality control.

The computer software that performs the process monitoring functions is likely to consist of several parts that reside in multiple computers. At the lowest level of the monitoring heirarchy, a data collection or acquisition operation is utilized to determine the status of the appropriate process parameters. This information is contained within the digital or analog signals that are produced by the process monitoring sensors. These sensors may be as simple as an analog or digital output voltmeter or as complex as a computer-based transducer that uses a self-contained computer program to convert a measured parameter into a calibrated, error-corrected voltage or current output signal. Figure 9.1 shows a LVDT position transducer that has been coupled with a computer to provide an improved measurement device. This unit physically appears to be a conventional displacement sensor, yet through the use of the internal calibration software it is able to achieve an extended range of travel while maintaining controlled linearity [1]. The improvement in performance is obtained by preparing a calibration table of the uncorrected LVDT output versus actual displacement. This measurement correction information is obtained using a calibration procedure that employs a high

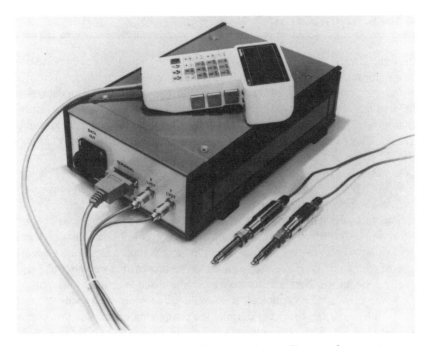

Figure 9.1   Computer-based position transducer (From reference 1.).

accuracy position measuring instrument such as a laser interferometer. Then, the appropriate correction factors are loaded into the computer memory for rapid access during the measurement process. During operation, the transducer calibration software is utilized to provide real-time correction of the LVDT output so that an extended range of travel is achieved without the degradation of system accuracy that is inherent with conventional LVDTs.

The previously described software is required only to input the raw data from the LVDT, obtain the corrected output value from a calibration table and provide this information to a position readout indicator. However, other types of transducers may require more complex calculations to provide corrected data. For example, a laser interferometer is sensitive to variations in atmospheric temperature, pressure and humidity. In this instance, a computer system that is used to provide a corrected output signal would require more input information than just the raw sensor output data to be able to perform the calculations necessary to determine the appropriate correction factor for the changing operating conditions. In addition,the mathematical operations required to obtain the corrected output information may be significantly more complex than simply selecting the correct answer from a calibration table. In this case, a slight time delay may be introduced into the measurement system due to the time required by the computer to execute the more involved calibration instructions. However, this time delay will be insignificant in most instances.

The data acquisition system that is used in the monitoring process also may employ a separate computer. In this situation, the data collection computer can be utilized to sample the various signals at different intervals. This asynchronous sampling is based on the expected rate of change of each particular parameter and avoids collecting excessive amounts of information for slowly changing signals while maintaining the responsiveness that is required for the shorter time-constant data. In a modern data acquisition system, the control software permits the performance of thousands of measurements per second with a resolution of better than one part per million. This level of performance is sufficient for monitoring most signals in the manufacturing environment. In addition, ample software is available through commercial sources to link the data acquisition system to a personal computer. In some situations, this capability permits added operational flexibility as well as allowing the data analysis task to be accomplished more readily.

An inherent feature in the design of CNC systems is the ability to monitor a number of process parameters such as the following errors within the servo systems, the status of control switches and limit switches, input data parity and the validity of the syntax in a part program command. While this real-time process data may not always be directly correlated with workpiece quality, it is directly related to the condition of the machine control system. In addition, it is a measure of process quality in the sense that various system malfunctions may be immediately detected. Also, other production information such as environmental data or on-machine gaging results may be recorded through special part-program commands.

If a higher level computer is acting as a DNC host to the CNC system, then it is likely that this host computer will maintain the data base of monitored information that records the condition of the manufacturing process. In this case, the local processor also will be involved in the synchronization of the data transfer to and from the host system. The software which coordinates the transmission of the manufacturing process data from the temporary buffer on the shop floor to the host computer must accumulate the monitored information in an orderly fashion, maintain a communications link to the host computer and transmit the data to the host at the appropriate time (before the capacity of the local buffer is exceeded). The DNC software that is being executed on the host computer must communicate with the shop floor processor, accept the incoming data and store it in the proper location for later retrieval and analysis. Once the monitored process parameter information has been transmitted to the host processor a data base system is likely to be used for further data manipulations. This data base software permits the storage and retrieval of the records of the process parameter characteristics as well as report generation. Analysis of the monitored information may be accomplished using the capabilities of the data base software or it may be performed using special applications software that is tailored to the specific operation.

Data Base Software

Data base software packages are computer codes that are usually one of many potential programs that may be stored in computer memory for access by an operator. The execution of the data base program is enabled through the use of specific operating system commands. Once activated, the data base software offers the user a unique set of computer commands which permit the storage, retrieval and organization of information at a much faster rate than would be possible if the operation were attempted in a manual fashion. In addition, the data base software also may be utilized by other programs that are operating on the computer system so that the desired data manipulations can be implemented without human interaction.

While computers are very fast at manipulating data, it is still necessary to store the contents of the data base in an organized fashion so that the information can be utilized at a later date. Just as with a filing system for paper documents, a coherent technique is required to keep track of all of the data. As an example, a data base used to characterize a manufacturing operation might record information on the size of a fabricated product, the completion date and the specific machine used to perform a crucial machining step. If all of this data were lumped together in the same data file without any way of distinguishing one piece of data from another then the data base would be of little value. In this situation, the only way of determining the size range for all parts machined for a particular month on a given machine would be to print the entire data base and manually sort through the printout to group the data into the three categories before trying to figure out what size values went with which dates and machine categories. (This is assuming that a user could determine which three pieces of data belonged together for a given manufacturing step.) In effect, the user would be overwhelmed with data that contained useful information. However, no convenient technique would exist for extracting the small amount of information that was pertinent to a particular problem from all of the data that described the history of the operation.

Computer data bases solve the problem of data organization and identification through the use of labels that are assigned to the particular categories of parameters being manipulated. In addition, through a linking process involving these labels, all of the related information for a particular entity is tied together even though it is not physically stored together within the computer memory. This means that once a particular entity (such as a part identity or machine number) is selected through a sorting process, all of the related information about the entity is also available. For example, it would be possible with a particular data base configuration to request that the system identify the names of all individuals that had access to a certain computer system on July 3, 1973. Then, using this list of user identities, it might be desirable to extract from the data base all the available information that has been recorded concerning these individuals (such as supervisor, job title, office number, the time of day that the computer system was utilized, and the files that were accessed). Of course, the availability of a given category of information is dependent on the structure of the data base that is programmed into the computer system. While data base systems permit the user to tailor the system operation to each specific problem, it is not possible to retrieve information that was never input into the data base. Nor is it possible to use relationships between data elements that were not defined in the beginning without reorganizing the data base. However, if it is recognized at a later date that the configuration of a data base should be different, it is usually possible to alter the original design to incorporate the needed changes without having to re-enter all of the previous data. This task generally can be accomplished using the inherent capabilties of the data base system, although the procedure must be done carefully to assure that data integrity is maintained.

One data base concept that has received a lot of attention is the relational data base. This type of data base is designed to provide quick access to user-specified subsets of data by maintaining sorted lists of key items that provide pointers into the remainder of the data base. While this does provide very quick access to the keyed features, it also introduces some additional system resource overhead. This is because additional memory and processor time are required to support this characteristic. Also, for a large data base this feature can be self-defeating if a user should decide to key all of the attributes. The better approach is to key only the more important attributes so that a rapid response is achieved without over burdening the system.

Data base queries may be conducted automatically from a program that is operating on the computer system or they may be performed manually during an interactive session at a computer terminal. In either case, the interactions with the data base are conducted using a special set of commands and syntax that are unique to the particular data base system (although most data bases have a somewhat generic set of commands). These commands accomplish tasks such as locating data (FIND command), sorting information based on attribute values (SORT command), performing logical and boolean operations on the data (AND, OR, EQUAL, etc., commands), and outputting results to the user (LIST or PRINT commands) [2]. Manual interactions with a data base are very useful when attempting to locate items for a special purpose search, such as determining a checking account balance when given a particular account number, or when attempting to locate a characteristic of a data set that has not been previously recognized. For tasks which recur periodically, such as issuing monthly status reports, it is more convenient to prepare a program which will run automatically to generate the needed report without requiring manual interaction by an operator. Another advantage to this mode of operation is that this type of activity can be conducted during whatever low usage periods exist for the particular computer system.

It is sometimes desirable to characterize automatically, in real time, a manufacturing process. In this situation, the process monitoring software extracts the information that describes the status of the operation from the system sensors, performs any necessary preprocessing and transmits process data to computer memory for storage. Additional information also may be input through computer peripherals such as a bar code reader or a manual data entry keyboard. Then, the data base system controls the storage and retrieval operations and may interact with other special applications software as a part of the data analysis process.

As mentioned previously, the software used to direct the operation of the data base consists of unique commands that must be assembled by the data base operator or other qualified personnel since this expertise is not usually available from individuals who have a manufacturing background. Execution of these application-specific commands causes the computer to perform the actions that are necessary to support the particular manufacturing operations. A sample operation might be to print out all of the part numbers for the out of tolerance workpieces that were manufactured over a period of one month and to list the shift on which they were fabricated. This information does not exist in any particular order in the computer but the data under a date label can be sorted according to the time period of interest. Additional sorting would be performed based on the quality attributes of the parts of interest and the results would be placed into a temporary storage buffer to simplify later operations (it is inefficient to perform the basic sorting operation each time another feature of this specific group of parts is examined). Next, it might be desirable to group the parts according to some attributes such as the amount of deviation from a particular tolerance dimension, the time of day that the finish machining operation began and the temperature of the machining coolant. In addition, this information is readily available for the preparation of summary reports, the assessment of the process status just before and after a malfunction occurred, and for use by process analysis software that is employed to provide a more complete assessment of the particular manufacturing operation.

Expanding the focus of the use for a data base management system, it becomes apparent that the information in one or more data bases may be of value to multiple organizations within a manufacturing facility. While there is some process data (such as the position offsets used to machine a certain workpiece) that is only important to the control of a particular machine, this information is also of interest in relation to the other machines in a shop or a plant since some of the other machines are used to perform a similar function. In addition, part program commands are a type of data that might be prepared on one computer system by one organization and then transferred to a data base on a manufacturing organization's computer for eventual down loading to a machine control unit. A similar situation also exists for other multiple use data that is needed for scheduling, production control, inventory control, planning and so on. The combination of these data bases into an integrated data management system is required to receive the full benefits available from the technology. Previously, a major road block to this integration of data was the difficulty in communicating between computer systems. With the improved computer networks and communications software packages that are currently available it is now possible to cross these physical boundaries. The result is that the user's access to manufacturing process information need not be limited by an individual computer system. In addition, it is neither necessary nor advantageous to maintain one huge data base, with all the process data, on a single computer. However, this distributed data base approach does involve additional complexity when updates are being made since some duplication of information is necessary for efficient operation. In addition, the control of access to the master files is crucial. The ability to read the information may be generated to a wide variety of users but editing, or modification, privileges should be tightly controlled.

Process Analysis Software

This software is used by a computer system to perform the real-time process characterization operations as well as to generate appropriate summary reports. The real-time analysis task primarily consists of the processing of information such as on-machine gaging data and status characteristics, as well as the screening of the information from the system sensors to determine quickly if the operation is functioning correctly. The acceptable range of values for these monitored parameters may be established through an intuitive process such as the application of engineering judgment or a statistical approach, such as the use of control charts, may be employed to provide a calculated basis for a quality assessement. One advantage to an acceptable range estimate based on a statistical process is that is can be automatically updated on the computer. In addition, the statistically estimated value is relatively free of personal bias when compared with intuitive techniques. Also, it provides an easily defined approach for determining the numerical value of a process characteristic.

If control charts are employed to estimate an acceptable range of values for the process parameters, it may be desirable to maintain this data on a host computer. However, the control limits for the pertinent parameters need to be readily available to the software that does the data screening prior to the transmission of the process information to the data base. This technique of maintaining the parameter control limits on the shop floor computer also relieves the burden of constant communications with the host system. In addition, it permits local shop floor operations to continue for an extended period in the event that communications with the host computer system are interrupted for some reason.

Summary reports are available in a variety of forms with most data base systems. These report writing packages can be tailored by the user for the specific needs of a given application. The output generally consists of printed data that is organized in the format defined by the user. Graphical representations of data are also available through other software packages. Figure 9.2 shows one graphical method of presenting data for a group of parts. With this technique the results for each certification dimension on a part type are plotted on a graph along with the mean and three standard deviations of the particular characteristic. The user-generated software needed to obtain this data is not difficult to write because the statistical information can be readily calculated using the internal data processing capabilities of most data bases. Also, the additional software that is needed to provide the graphical output is relatively easy to acquire.

Figure 9.3 shows another graphical approach for presenting quality information about a manufacturing process. This technique utilizes the statistical control chart approach that was discussed in an earlier chapter. Commercial software is readily available to maintain a process control chart. However, user intervention may be required in the detection of process shifts and error conditions as well as the recalculation of the control limits. When it is necessary to automate these actions then it will be necessary to acquire additional user-specific software which addresses this specific task. A variety of other process analysis software is available to perform a wide range of engineering and scientific calculations and these programs are available for use on the computer that is maintaining the data base. The only difficulty that may be encountered in using this classification of software is the interface to the data base.

Error Compensation Software

In conjunction with process monitoring and analysis it is frequently necessary to compensate for process errors until such time as the error source can be eliminated. On a machine tool these errors generally exhibit two types of forms. The simplest type of problem to accommodate is one which causes a simple offset error in one or more of the machine's axes. In this case, it is only necessary to recognize the error (which is sometimes a nontrivial task) and implement a single offsetting displacement of the appropriate axis to obtain the necessary corrective action. The software that accomplishes this task is relatively simple because only a one-time position adjustment is required and this is easy to execute.

A more complex situation exists in the case in which the error source causes a continuously varying position error on one or more axes. This is the type of error condition that can not be corrected with a simple axis offset at a single point in the machining operation. In this instance, it is necessary to determine the effects of the error source ahead of time and make continuous corrections throughout the machining operation. Alternatively, in some situations it is possible to implement a closed-loop error correction system that operates in real time to continuously adjust the machine's operation in response to the measured error conditions.

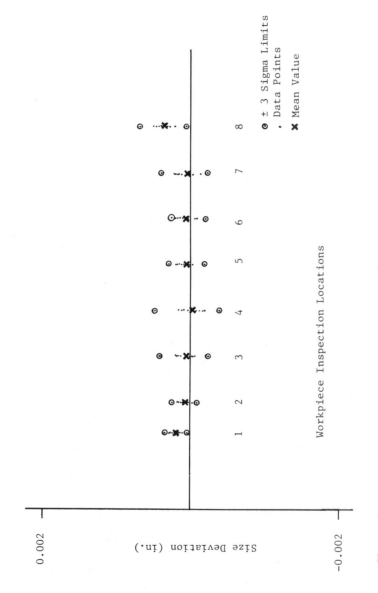

Figure 9.2 Graphical presentation of workpiece quality.

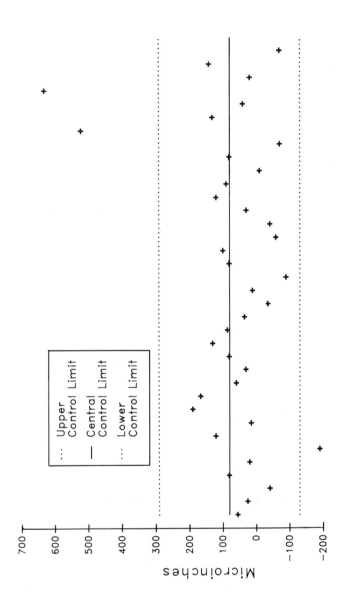

Figure 9.3 Control chart presentation of workpiece quality.

One example of the first type of error condition is the problem of machining a workpiece on a multi-axis milling machine when the exact location of the part relative to the rotating cutting tool is unknown. This information is needed prior to the beginning of the first machining pass to allow the determination of the amount of material that will be removed from the workpiece. In order to obtain this information, a simple probing cycle (using either a gaging probe or merely toughing up with the cutting tool) can be used to establish the location of a datum surface on the part relative to the machine's datums. Then, this position information can be utilized to establish the appropriate machine offset. In this type of situation, it might be argued that it is easier to perform this task manually than it is to provide the hardware and software required to meet the objective. However, the automation benefits that are provided by this feature make the effort very worthwhile.

Another example of employing a simple axis offset to compensate for a machine error source is demonstrated by the leadscrew compensation feature available on most CNC systems. In this situation, continuously changing position errors exist along the length of the slide travel due to the mechanical inaccuracies of the leadscrew. The resulting axis displacement errors are characterized using a very accurate position measuring device, such as a laser interferometer. The positioning errors are recorded at incremental positions along the length of the slide travel and stored in an offset table in the control system memory. During operation, the error data is extracted from the offset table by the CNC executive program and used to correct the position of the machine's slides as movement occurs past each of the predefined incremental locations.

The on-machine inspection of key workpiece features for offset calculation is another instance in which the actions required to correct the action of the machine's motions are relatively simple. In this case, key dimensions such as a diameter or a length (which can be corrected with a simple axis offset) are measured prior to the last machining pass. The inspection results are compared by the CNC executive program with target values that are associated with the specific part. Deviations from these key dimensions are utilized to calculate machine offsets that are automatically incorporated to achieve the desired feature sizes. The software required to accomplish this task is well within the capabilities of modern CNC systems although it may not exist as a standard system feature.

A more complex problem is the case in which the needed axis position-correction values are a function of the positions of all of the machine's axes. In this instance, a complete error map is needed that provides the axes correction information for each point within the work zone of the machine. In addition, this data may be influenced by the temperature of key locations on the machine so that the correction tables and/or algorithms may become quite complex. On-machine inspection of workpiece contours and the adjustment of the tool path based on this information is another difficult task. One approach is to utilize the inspection information to prepare a revised part program that is based on the existing contour errors; however, this is a fairly complicated task and must be performed on a host system. An alternative is to utilize the computational abilities of a CNC to generate an error table that is based on discrete inspection points and use a form of cutter compensation software to handle the implementation of the appropriate offset for each block in the machining part program.

In developing a systematic and generalized approach for determining the total error of machine tools, all the factors which affect the accuracy of the final position of the cutting tool with respect to the part must be considered [5]. Such an approach offers the opportunity of implementing a total error compensation system. To achieve this objective, the NBS has generated a mathematical error model through the use of homogeneous transformation matrix manipulations and the assumption of rigid body kinematics. The approach is to utilize predictive machine calibration techniques that accommodate each of the machine error components. To obtain this predictive model, the machine's motions are calibrated while allowing the key process parameters to vary over the range of likely operating conditions (slide positions, temperature, workpiece weights, etc.). Once the individual error relationships are established, then the resultant error vector can be calculated in real time for a given set of operating conditions. The final step is to compensate for the composite positioning error of the cutting tool with respect to the workpiece.

The NBS has developed this real-time compensation system into a peripheral for a machine controller. This system monitors the condition of the appropriate process parameters and injects an error compensation signal into the machine control circuitry as shown in Figure 9.4. The software in the error compensation

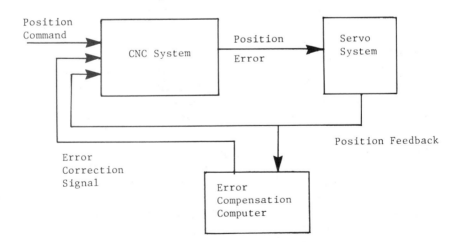

Figure 9.4   Block diagram for error compensated CNC system axis drive (from reference 3.).

microcomputer runs in synchronization with the machine tool control system and receives position feedback information at the CNC system's position update rate. Based on the current machine axes positions, the direction of motion, the temperature data and tool setting information, the microcomputer software calculates the current error vector and the corresponding servo compensation needed to correct the position of the machine's axes. Then, this information is transmitted to the CNC system as additional axes following errors.

This software was written in the Programming Language for Microcomputers (PLM) which is a high-level structured language. This allows easy modification of the error model coefficients whenever necessary without changing the main body of the software.

An advantage to this software error compensation technique is that once it is set up it runs continuously without requiring addition outside computing resources. In addition, it does not interfere in anyway with the basic servo control algorithm of the CNC system. A disadvantage of this approach is that a detailed knowledge of the CNC system is required to obtain the necessary position feedback information and to inject the additional following error data into the servo system.

An alternate software compensation approach is to characterize the workpiece errors at an appropriate point in the fabrication cycle and generate a modified part program to correct the existing errors. The advantage of this approach is that the compensation values are tailored to the specific errors on the part. Disadvantages associated with this technique are that the errors must remain constant between the characterization and correction cycles and additional external software computations are required for each corrected part program.

This discussion highlights two main approaches for applying software error compensation techniques. The choice for the system designer is that the correction software can be executed locally or at a remote location. The advantage to the remote calculation method is that the computing resources are generally much greater. The disadvantage is that these resources may not always be available. As might be expected, the tradeoffs associated with each particular application must be considered to determine which is the best approach to pursue.

Modeling Software

One use for this category of software is to employ it in the evaluation of the suitability of certain process elements such as the NC part program. Conventional methods for checking the correctness of NC part programs have depended on techniques such as "dry running" the program and the use of a NC plotter to "sketch" the tool path defined by the machine motion commands. Unfortunately, these two techniques only provide an estimate of the part program quality and an accurate evaluation of the program performance does not occur until an actual workpiece is machined. Today, CAD/CAM software offers an alternative approach. This technique permits the user to input the definition of a desired work-piece configuration into a computer system, generate the required part-program commands needed to fabricate the final product and automatically inspect the cal-culated tool path commands to be sure that the machining operation will provide the intended workpiece configuration. In addition, this information may be dis-played on a color graphics terminal and viewed from different perspectives. Also, a zooming capability exists that allows the user to obtain a close examination of different segments of the tool path and the final workpiece.

Another reason why it is useful to model a manufacturing operation is that this procedure results in a detailed understanding of the process errors as well as the manner in which they interact. The result is a mathematical relationship that corre-lates the status of the various process error sources and the quality of the final product. With this manufacturing process model comes the ability to estimate the likely effects of an excursion in the various error conditions that could occur during the manufacturing process. As mentioned previously, the use of this process model during the manufacturing operations enables the estimation of product qual-ity at a point in the fabrication cycle at which it may still be possible to make system adjustments to accommodate variations in erratic process parameters.

Figure 9.5 shows a cross-section view of a lightweight mirror that is attached to a machining fixture in preparation for diamond turning the flat face of the work-piece [3]. Because of the tight tolerances associated with this product, it was neces-sary to consider the part-deformation effects that would be introduced by the centrifugal forces associated with the turning operation. Since the part would be elastically deformed and then machined, the "spring back" of the material when the work spindle was halted would automatically induce errors in the workpiece. Figure 9.6 shows the system deformation predicted by a finite-element analysis of the machining operation. This software modeling technique permitted the preliminary evaluation of multiple fixturing options without having to perform extensive hardware tests.

Figure 9.7 shows a photograph of a machine tool that was plagued with servo system instability problems at the microinch positioning resolution that was desired for a particular task. In this situation, the undesirable operating conditions were analyzed using structural vibration data [4]. The axes positioning difficulties were traced to a sharp resonance that existed in the cross-slide servo system at a fre-quency of 38 Hz. Use of a digital signal analyzer and a commercial modal analy-sis software package lead to the discovery that the first vibrational mode of the structure also was at 38 Hz. Figure 9.8 shows the machine tool structural dis-placements at this mode. These deflections acted as a direct disturbance to the slide drive system. Since this disturbance occurred at a frequency that was within the desired operating range of the servo system it was not practical to construct an electronic filter to modify the system performance. Instead, the modal analy-sis process was continued in an attempt to eliminate the source of the disturbance.

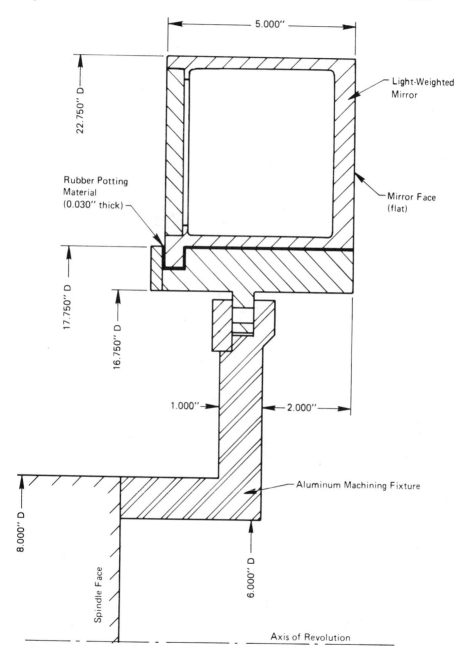

Figure 9.5 Cross-section view of workpiece manufactured through diamond machining (from refence 4).

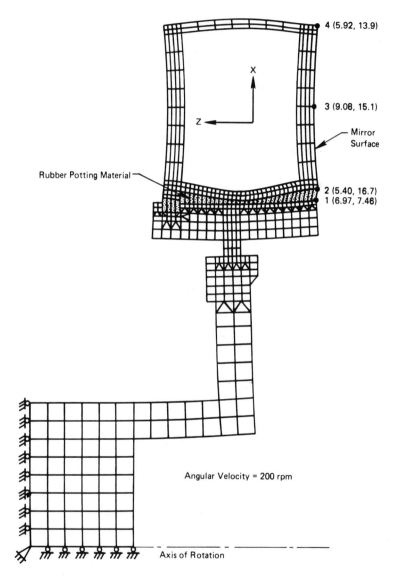

Figure 9.6 Finite element prediction of workpiece deflection during diamond machining operation (from reference 4).

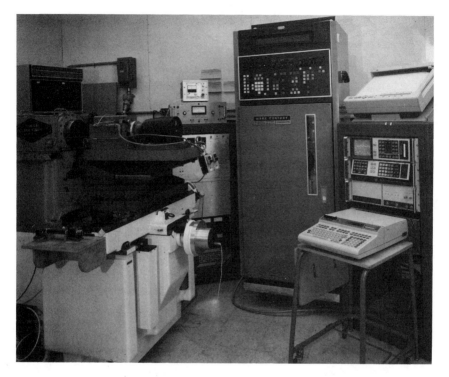

Figure 9.7 Machine tool used for vibration analysis (from reference 5).

A subsequent structural modification of the machine's support platform was used to solve the problem of the servo system instability.

Both of the two previous examples demonstrate instances in which computer software provided an analysis technique that enabled critical manufacturing system errors to be characterized. In addition, the software also provided a simulation tool for use in the error correction phase of the activity that avoided extensive fabrication of test hardware. Unfortunately, the difficulty that is encountered in designing a system to perform in-process monitoring and error correction is that the excursions in the manufacturing process parameters must be detected and corrected (or an error flag generated) while the manufacturing operation is in progress. In order to accomplish this, a system model is needed that utilizes input from the key process parameters and analyzes the effects of all of the error sources to determine the status of the operation.

Control chart techniques are an acceptable method of showing that individual parameters are operating within desirable limits but this method does not readily address the problem of the interaction of multiple parameters. This approach will

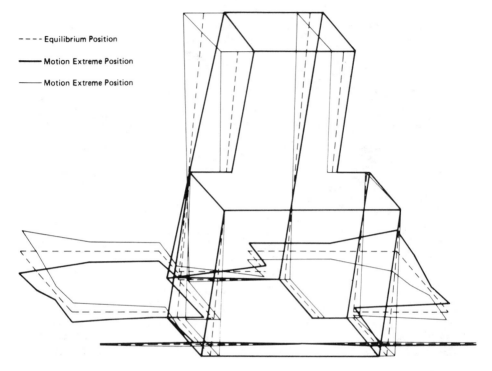

- - - - Equilibrium Position

——— Motion Extreme Position

——— Motion Extreme Position

Figure 9.8 Structure displacements that occurred in machine tool (from reference 5).

flag the existence of unusual variations in one or more process variables but it cannot predict the cumulative result of the variations. Stated another way, statistical control charts are capable of detecting that a part feature or other manufacturing process variable is drifting out of tolerance. However, this approach is unable to predict the manner in which multiple error sources are likely to combine to cause a particular process deviation. A process model that accurately predicts the process output in response to the operating conditions is required to accomplish this task. On the other hand, control chart techniques are adequate for modeling many processes because these operations are likely to experience excursions in only one parameter at a time. In addition, if multiple parameter excursions do occur, at least the control chart technique will provide an early indication that something is wrong with the manufacturing operation. However, an additional shortcoming is that in many cases the process model approach also provides information that can be used for error correction. The control chart can only indicate that a problem exists, it has no way of providing the information needed to adjust for the variation in the operating conditions.

References

1. J. V. Moskaitis and D. S. Blomquist, A microprocessor-based technique for transducer linearization, *Precision Engineering*, 5:1 (1983).
2. W. E. Barkman et al., Using A Data-Base Management System To Characterize A Manufacturing Process, Martin Marietta Energy Systems, Oak Ridge Y-12 Plant, Oak Ridge, Tennessee Y-2291, October (1984).
3. M. A. Donmez et al., A general methodology for machine tool accuracy enhancement by error compensation, *Precision Engineering*, 8:4 (1986).
4. J. E. Stoneking and H. L. Gerth, Mechanical Deflection Analysis of Diamond-Turned Reflective Optics, Union Carbide Corporation-Nuclear Division, Oak Ridge Y-12 Plant, Oak Ridge, Tennessee Y-2108, December (1977).
5. S. S. Douglass et al., Control Systems Improvements in a Precision Coordinate Measuring Machine, Union Carbide Corporation-Nuclear Division, Oak Ridge Y-12 Plant, Oak Ridge, Tennessee Y-2249, September (1981).

CHAPTER 10

System Integration

Introduction

System integration, and its intimate involvement with manufacturing hardware, software and personnel, is the point at which all of the various planning activities come together and the system is made to work to meet the original goals of a project. This implementation step occurs in a high profile environment because everyone can watch equipment being installed and begin operating (or malfunctioning). The success of the initial installation is of major importance to the eventual success of the complete manufacturing system. Smooth implementations tend to lead to operations with successful manufacturing quality control, while poorly performed system integrations invariably leave the system user with a high priced eyesore that is an embarrassment to everyone.

Quality software that performs the desired tasks is one of the key factors in achieving a successful system integration. Hardware is generally very visible in a physical sense, while software is something that "runs on the computer" and does not necessarily receive sufficient attention until it is time to depend on it and it does not perform as expected or as required. This mismatch between hardware and software is very detrimental to the manufacturing system integration process. Without the required degree of integration the most that can be realized with the manufacturing operation is the implementation of islands of technology.

These islands may be very sophisticated and capable of wonderful state-of-the-art accomplishments. However, without complete system integration, there will be remaining operational bottlenecks that preclude the realization of the manufacturing system improvements that should otherwise be available. In addition, the overall process quality and productivity may not be significantly altered unless the integration process is complete.

Integration involves the act or process of uniting or forming the pieces of a system into a complete unit that functions in a coordinated fashion to perform some task or to achieve an objective that could not be readily accomplished otherwise. There are, of course, various degrees of success or completeness to each integration task. In addition, a successful integration project does not imply total automation of a given system. In some instances, there may be operations or procedures for which it is more appropriate that a manual or semi-automatic approach be implemented in order to obtain the best solution to an overall objective.

In the manufacturing arena, integration operations frequently involve subsystems that may be categorized in general terms as having characteristics that are predominantly mechanical, electrical, hydraulic, and so on. In addition, each of these subsystems also may require a significant degree of integration to achieve the needed capability. At some basic level the process begins with the appropriate building blocks of hardware, software, and so on that are assembled into increasingly more complex units as the process continues.

Bringing all of the particular system's various pieces together into a complete, successfully functioning module requires that the necessary degree of interactions or communications be established between the different pieces of the puzzle. This interaction between entities is required among certain elements at a given level of the system as well as between the various levels of the system. All of the various subsystems contained within a given portion of the overall system hierarchy must be inherently compatible with each of the other units with which they must interact. Alternatively, a suitable interface must exist to allow interactions between these separate entities.

Examples of different types of interface devices include the coupling between a motor shaft and a machine leadscrew (mechanical), a computer-controlled magnetic relay (electrical) and a light emitting diode signal isolation module (optical-electrical). In each case, the interface device permits the necessary interaction between different types of entities; however, the manipulation involved is different from what occurs with a sensor. (Of course, it can be argued that a sensor is an interface between the physical phenomena and the control portions of a system; or conversely, that an interface is a sensor that monitors one characteristic and converts it into something to which another piece of equipment can respond.) In some applications, these parameters are similar in nature (such as motor rotation and leadscrew rotation), while in other instances, the parameters are significantly different (such as an electrical signal and a position displacement). In general, sensors are utilized to monitor process parameters, while interfaces are used to assure compatibility between the different elements of a process.

The interface process described above is similar to what is required to allow different computer systems as well as computer programs to interact with each other. However, one important difference that exists in this type of environment, as compared to what has been previously described, is that in a computer system the parameters that are being transferred across the various interfaces always can be characterized as information or data since this is all that a computer system requires. It does not need mechanical devices such as gears to interconnect its subsystems. In a sense, the elements of the system are so similar that only an information coversion is required. (A similar situation exists when a flexible coupling is used to interconnect two rotating shafts. No electrical interaction is required because only mechanical data is being handled.

In a digital computer, all the information is contained in binary data bits or zero's and one's. It is the meaning of the combinations of these information bits that varies among different computer systems—just as different combinations of letters are employed to mean the same thing in the English and Portuguese languages. The conversion of information from the type or format that is used in one computer system to the structure that is required by another system can be accomplished through a data translation process. This activity is carried out by acquiring data from one process and reconfiguring it to match the format that is required by another process, which is analogous to what is done when an interface device is utilized to integrate a mechanical and an electrical subsystem in a given application.

An example of the rewards that can be achieved through the successful integration of the various subsystems in a manufacturing operation is demonstrated by General Motors plants in Linden, New Jersey and Wilmington, Delaware [1]. At these facilities, equipment such as automatic guided vehicles, robots and in-process and postprocess inspection stations are integrated into a manufacturing system that has eliminated many operations that were previously performed by hand. The result has been improved quality, reduced manufacturing cost per vehicle and the potential to double productivity.

This chapter discusses the blending of the various facets of process quality control, as presented in the earlier chapters, into an integrated manufacturing system that is capable of providing high-quality products in a timely and cost-effective manner.

Hardware Integration

The subject of hardware integration involves not only the blending together of existing hardware components into a viable package, but also the enhancement of existing systems through the interfacing of additional advanced technology hardware into a manufacturing system. The requirement for existing hardware integration is encountered when it is necessary to merge currently available equipment into a larger system that performs a desired task in a better manner. This does not necessarily mean that a high degree of automation is employed to eliminate all manual tasks from a manufacturing operation. In fact, that approach may not even be a viable method to achieve the project goals. Instead, successful hardware integration means that each of the hardware elements within a system is able to perform its assigned tasks as intended and it is able to interact with the other elements of the system without encountering any difficulties. The result is that the manufacturing system functions in such fashion that the flow of product through the process is smooth and well controlled and the quality and productivity of the operation are sufficient to meet the requirements for the particular product.

The effort required to perform a given hardware integration task can be as simple as matching the appropriate sensor to a particular production operation. This only requires selecting the best measurement transducer for the given task so that a correct assessment can be made of an operation's status. In contrast, the integration effort also may be as complex as combining a group of machines and part handling equipment together in an automated cell in order to achieve totally untended manufacturing operations. This type of project requires many individual system elements and interconnections between a wide variety of equipment so that a considerable effort is required in the planning as well as implementation stages.

A similar situation also exists when integrating new equipment into an existing manufacturing process. In both cases it is necessary that each item of equipment perform its intended operation in the desired manner, and also interface with other equipment. The requirement for a hardware interface with other equipment means that both the mechanical and electrical aspects of the problem must be addressed. The various machines must be physically capable of performing the basic job requirements as well as the necessary interaction tasks. In addition, any electrical or other junctions between the different entities must be of the appropriate type. A tool changer that has insufficient tool storage capacity, a robot with insufficient lifting capability and a machine with insufficient slide travel are all examples of situations that lead to mechanical integration difficulties. While the individual piece of equipment is functioning properly with respect to its original design, it is inappropriate for the current application. An example of an electrical interface problem is a device has a self-contained output signal display but no auxiliary output for use with other equipment. In this case, the device performs its basic task but it requires manual intervention to interpret the output signal. An instance of an integration problem among digital computers is when different systems do not use the same communications protocol. Both computer systems may be capable of performing the required individual tasks but the exchange of information between the systems is unable to occur without additional special purpose equipment.

To someone with relatively little systems experience it may seem that one viable approach to upgrading a manufacturing operation is to "go in and stick a computer on everything." Then, when the computer is installed it will just be a matter of letting it solve all the problems inherent with the existing equipment, via compensation algorithms, as well as perform the desired control tasks. (Instant computer-integrated manufacturing!) Unfortunately, while computers are very predictable and repeatable, they are also remarkably lacking in creativity and intelligence. They only do what someone has programmed them to do, which is not necessarily the same thing as what is desired for them to do.

It is not appropriate to attempt to utilize a computer to correct machine problems that also could be rectified with a moderate degree of effort. While computers can perform complex error correction operations, this usually requires extensive control of reference datums and operating conditions to be sure that the appropriate corrections are being implemented for a given set of operating conditions. Although this may be necessary in some applications, this approach frequently introduces unnecessary complexity and time delays to an operation.

Successful hardware integration is usually achieved by employing the most appropriate hardware for the task at hand. If the hardware is the weak link in a system, then software corrections often can be depended on to compensate for these problems. However, it can turn out that the software development costs are as large or larger than the expense that would have been incurred if the approach had been taken to fix the existing problem in the first place. Of course, this software may have applications in other areas so that the development expense is shared by multiple projects. On the other hand, there is also an ongoing expense associated with the use of software compensation techniques as well as the cost of maintaining the software so that the best solution is not necessarily the same in all instances. The approach taken in a given situation must be choosen in light of the various factors associated with the particular manufacturing operation.

In one sense, the hardware integration problem is the easy task and the real difficulty lies in the software integration. This statement is not intended to imply that hardware integration is trivial; only that in many instances, commercial equipment is readily available to meet the hardware requirements of a manufacturing system. However, the availability of commercial software to "tie everything together" is rather limited. This is partly due to the rapidly evolving capabilities of computer control systems and also due to the nonuniformity among the varied applications. In many situations, a particular application requires a significant amount of custom software in relation to what a vendor has provided in other installations. However, given a clear description of a computing problem, there is not much that a talented, motivated programmer cannot accomplish. The key to the success of this custom software is the accuracy and completeness of the software specification, as well as ongoing communications between the vendor and customer so that both parties have a sound understanding of the capabilities and idiosyncrasies of the eventual system.

Figure 10.1 shows a set of specifications for a precision flexible manufacturing system (FMS) for turning operations that was studied by the Oak Ridge Y-12 Plant in the early 1980s. A completely automated system was desired that would produce stainless steel hemishell workpieces to a tolerance of plus or minus 0.0005 in. (12.5 microns) on contour and plus or minus 0.001 in. (25 microns) on wall thickness. In addition, the system was to achieve these tolerances to a "5 sigma" level of confidence (the feature tolerance is equal to 5 times the standard deviation of the process) without requiring operator interaction. Figure 10.2 defines the sequence of the various manufacturing operations that were required to fabricate the workpieces.

As a preliminary planning step, a feasibility study was performed by several companies, two of which were machine tool manufacturers. The general consensus was that both state-of-the-art equipment and software would be required to obtain the stated objective. It was felt that the system tolerances could be met by the basic machining hardware but that there were uncontrollable parameters

        The following material provides a functional
description of a Precision Flexible Manufacturing System
that shall be capable of producing uranium hemishell
workpieces that range in size from 4" to 10" in diameter and
up to 1" in wall thickness. The parts must be machined to a
size tolerance of +/- 0.0005" on contour and secondary
features such as steps and chamfers and to within +/- 0.001"
on wall thickness.

The machine configuration shall be identical to the Plant's
existing family of precision t-base machine tools.

The system shall be capable of sustaining a 3 to 5 sigma
process within the stated tolerances for the described
parts.

The system shall be capable of machining either the inner or
outer contour last and of generating any turned secondary
features such as grooves, threads, etc.

The following system features shall be included:
          Automatic part/fixture exchange and alignment
          Automatic tool changing
          Automatic tool setting and edge condition sensing
          Automatic tool height control
          Compatability with existing DNC system
          Automatic chip breaking and removal

The system shall function in a deterministic manufacturing
mode wherein automatic process monitoring is employed to
produce the desired product output.

Figure 10.1 Specifications for a precision flexible manufacturing system for turning operations.

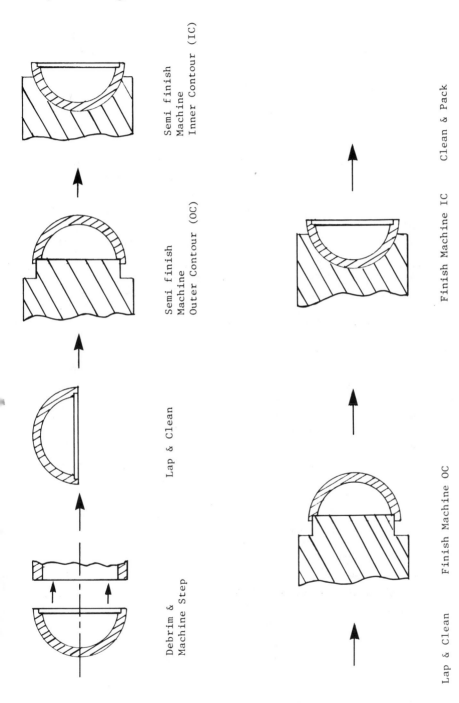

Figure 10.2 Sequence of operations for fabrication of precision hemishells.

such as tool wear which would introduce an unacceptable level of variability into the process. Elaborate in-process monitoring and correction procedures were proposed to obtain the desired tolerance levels for the complete manufacturing system. In addition, the magnitude of the integration task was such that it was only possible to roughly estimate this expense because no one in the machine tool industry had ever done a project of this complexity before. The financial implications of this uncertainty were sufficient to elevate the price of the project to a level which could not be justified. As a result, the total project was not funded, although portions of the problem did receive further attention. Eventually, it was decided to pursue a smaller incremental step and acquire an enhanced retrofit package for the existing precision lathes which would provide the needed workpiece tolerances but only in a semi-automatic mode of operation. However, even this scaled down version of the original task contained a significant degree of uncertainty concerning the requirements to provide complete system integration.

Software Integration

Software integration implications are encountered in almost all phases of the integration of a manufacturing system. Although software considerations are not required to assure that two mechanical parts fit together correctly, software integration activities are frequently required to control the operation and interaction of the system's mechanical components. At the lowest level of software integration in a CNC system, the executive software is responsible for integrating the commands in the part program with the status of the machine to produce the desired result. A good example of this activity is the closed-loop position control used on a machine's axes. In this instance, interpolation software is utilized to generate a series of desired position commands that are based on the axes departure and velocity commands which have been read from the part program. Also, position feedback software is used to convert the information from a position transducer to a displacement value that indicates the incremental position of the machine's axes. Then, the control system software integrates the position command data and the true machine position information together to arrive at the actual command used to drive the axis servo amplifier. Figure 10.3 shows a block diagram representation of the way this process was coordinated in a control system that was used with a laser interferometer position feedback transducer.

On a larger scale, software also is utilized to coordinate the activities of the various entities in an automated manufacturing facility. Once the hardware is installed correctly and functioning properly it is the manufacturing system software that provides the coordination that makes the system operate in a manner such that it appears that "it knows what it is doing." Software integration can be a complex process because it is necessary to decide both what should be done in a wide variety of circumstances and also how to achieve the desired results.

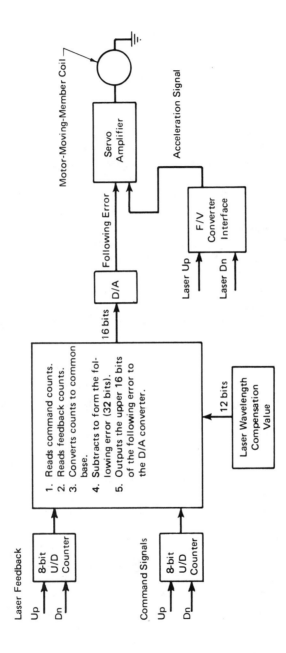

Figure 10.3 Block diagram representation of software integration of a machine servo system.

One way of visualizing the problem is that it is the task of the system hardware to perform the required mechanical functions while it is the responsibility of the system software to coordinate these mechanical activities as well as recognize and react to situations in which the hardware is not functioning in a desirable fashion. Generally speaking, hardware is not expected to perform the error correction tasks that software accomplishes. On the other hand, it is usually much easier to implement a software modification than it is to perform a hardware alteration (although if good documentation practices are not followed then significant problems can occur).

As mentioned in previous sections, it is desirable, for reasons of autonomy, to place as much "computing power" as is reasonably practical at the lower control levels of a manufacturing system. This control system configuration is required so that the shop floor equipment can function for extended periods of time without requiring communications with higher levels of the controls heirarchy. In addition, when tasks are partitioned in this manner, it is possible for the host system to function properly without having to receive information concerning the details of the individual operations. The host computer only needs to have information which defines the status of the individual jobs so that the previously scheduled sequence of manufacturing activities can be coordinated.

While the sequencing activities of the host computer are relatively straightforward during normal conditions, complex controls requirements can arise in the event of an unscheduled operations shutdown due to a system malfunction or emergency condition. Since this is an unplanned excursion in the system operation there is not likely to be an orderly termination of the work which is in progress within the system. In addition, performing a system restart is more complex than an initial start-up because provisions are needed to deal with the product that had not reached the end of the manufacturing operation at the time of the system shutdown. If a restart capability is not integrated into the operation, then it may be necessary to clear the manufacturing stream in a semi-automatic mode and then "bring the system up" as if it were starting with the beginning of another production run.

Another software integration issue is the utilization of the information that can be generated through the incorporation of various in-process monitoring transducers into the manufacturing process. At one end of the spectrum is the situation in which no operations data is collected and no attempt is made to manipulate the performance of the system based on real-time measurements of product quality. The other end of the spectrum may be described by the term "data pollution" which is the condition in which excessive amounts of process variable information is available with no consideration given to its utility. In this case, a frequently encountered condition is one in which it is impossible to perform meaningful process adjustments based on the feedback data, due to the volume of information that is produced. The problem is not knowing which of the sometimes conflicting data is correct and which information should be used as a basis for the process adjustments.

The middle ground for the in-process data collection/utilization operations is the point at which a system should be designed to function. Following this guideline, sufficient process information will be obtained in a timely manner so that the current status of the process can be evaluated. In addition, this real-time process information can be integrated into the control commands that are used to direct the operation of the system. This approach permits timely system control as discussed earlier in relation to the closed-loop position servo system. Alternatively, if this information is not integrated into the system's operation then the data is still valuable as a historical record. While this may be useful from a maintenance stand point or in a trend analysis it does not do much to assist the real-time in-process quality control efforts. If the data is not sufficiently reliable to be used for real-time process control then efforts should be expended to improve the quality of the data. If this is not practical, then the data collection/analysis operations should be halted. This is because it is a waste of resources to continue the information collection/analysis operations with unreliable process measurements. The "but its the best data that we have" syndrome ignores the fact that if the information is unreliable for control purposes then it is also inadequate for analysis activities.

In the ideal situation, the manufacturing system would be able to do more than just evaluate the current status of the process and determine the appropriate actions to achieve the desired level of process parameter control. In addition, it would also have the capability to evaluate the results of its controlling activities. Then, through a correlation of the expected process results with the actual results a process model would be constructed and continually updated for use in the estimation of the appropriate control parameters for a given set of operating conditions. While the software integration activities needed to achieve this level of manufacturing process control are quite formidable, this type of control algorithm is used in some existing learning systems. This type of manufacturing system could be accurately described as a completely integrated closed-loop manufacturing operation.

Integration of Statistical Methods

As mentioned in the earlier discussions concerning the application of statistics in manufacturing operations, a certain amount of error exists in all processes. This statistical error is independent of product tolerance. The error exists because all the output product is not identical, even though its pertinent characteristics may fall within the allowed manufacturing tolerances. Because of this lack of perfection, each of the quality features of a workpiece will have values that are distributed about a number that can be defined using a statistic called the mean or the expected value of the individual parameter. In addition, a tightly controlled operation will have process parameter variations that are characterized by small standard deviations, while an out-of-control process will exhibit excessive scatter. Another possibility is the occurrence of a tightly controlled process which may be operating around the wrong set point. In this case, the difficulty is that while good repeatability exists from one unit to the next, the quality features are still unacceptable. Fortunately, in this situation, it is usually possible to center the process through available adjustments so that the desired tolerances are achieved. As might be expected, this type of error condition usually is relatively easy to correct in comparison to the situation in which the average value of the product acceptance characteristics is acceptable but excessive variation exists between sequentially fabricated products.

The deviation between the actual and intended results of an operation is composed of disturbance parameter contributions from all the various error sources contained within the manufacturing system. These multiple parameter excursions interact in a complex manner wherein the results of some of the errors are additive and others act to cancel each other. Since the individual error sources are essentially independent of each other the composite error value is difficult to predict without knowing the exact characteristic of the various parameters at all times. However, it is not unrealistic to expect these error elements to behave in a statistically controlled fashion.

A realistic goal in a manufacturing operation is to attempt to keep these normal process changes within normal bounds. This allowable boundary range can be specified in relation to the statistical variations that would be expected to occur due to normal random events. The factor which determines whether or not these normal process variations are excessive is the manufacturing tolerance that is specified for a given operation or product. If the normal process variations do not result in a level of product quality that is unacceptable, then successful results can be achieved by maintaining the key process parameters within limits defined by statistical control charts. Also, in some situations a manufacturing operation may be exhibiting a lack of statistical control but still be producing product that meets the desired specifications. In this case, it may be desirable to schedule corrective maintenance activities at an appropriate time in order to keep the situation from continuing to degrade the product quality. However, immediate action is not necessarily justified unless there are other contributing factors involved.

If the normal variations in the process parameters produce products that are not acceptable, then the manufacturing operation must be improved. This improvement in output quality can be achieved through external actions or a "sift and sort" approach employed to cull the good ones from the bad ones. The problem with total dependence on postprocess inspection for determining product quality is that it only detects excursions in the process quality after the products have completed the fabrication cycle. In addition to this being an expensive and time-consuming approach in many instances, this technique does not provide much information about the cause of a malfunction. This operating procedure only verifies the effect of the process disturbance.

The integration of statistical methods into the manufacturing system avoids some of the expense associated with sift and sort operating procedures. One significant benefit occurs through the implementation of a statistical sampling plan for product certification. If the manufacturing operation is reasonably well controlled then it is not necessary to inspect every feature on every part to determine that the product is acceptable. Instead, statistical techniques can be utilized to provide assurance that the products meet the specified tolerances. Another advantage is that statistical analysis of real-time process parameter feedback measurements provides a more timely indication of the quality of the fabrication process. This early warning information permits the detection of unusual operating conditions within the process in conjunction with the support of the sampling for certification activities. In addition, parameter excursions are frequently detected before the problem becomes excessive and a method is available for pinpointing the cause of variations in the quality of the fabricated products. Also, the use of sound statistical techniques in the testing and characterization of a manufacturing operation leads to an accurate assessment of the process modifications that are needed to improve the overall quality of the system.

The deterministic manufacturing method of operation depends explicitly on the integration of statistical methods into the workpiece fabrication process. This technique for producing products assumes that manufacturing operations exhibit only cause and effect actions. Therefore, this operating procedure permits the status of the key process parameters to be utilized in real time to accurately represent the condition of the workpieces at the various stages of the manufacturing cycle.

In order to implement the deterministic manufacturing approach, it is first necessity to statistically establish the direct relationship between particular manufacturing process parameters and the quality of the manufactured product at the appropriate points throughout the processing cycle. Then, the status of these parameters can be monitored in real time to assure that everything is stable during the various manufacturing operations. This operational procedure avoids the necessity of having to certify the finished workpieces by means of a complete inspection of all features on all parts. Instead, the process is certified to have produced the workpieces without experiencing a larger than normal variation in any of the key process parameters.

Simple control charts of the key process characteristics will provide a statistically sound approach to tracking the system performance and will quickly detect excursions in the manufacturing process. However, this process monitoring method is not able to estimate the quality of a product based on the in-process measurements. It is only able to estimate statistically that the product characteristics fall within a certain range of values. To estimate the actual values of the process characteristics requires a process model which is often more difficult to acquire. Another advantage of the use of process models is that this technique can deal with excursions in multiple process parameters and estimate the resulting product quality.

Both control chart and process model techniques share the advantage of offering the rapid detection of potential problems and the elimination of the requirement for extensive postprocess inspection. However, even with these approaches, some certification capability must be retained to verify the validity of the system and to provide more extensive information about those products that experience unusual operating conditions during the fabrication cycle. These questionable workpieces may be acceptable, but additional certification measurement steps are required to determine the final characteristics of the parts.

Facility Integration

The integration of new or modified equipment and processes into an existing manufacturing facility is not necessarily a routine task. While these changes may be implemented as a part of a quality improvement program, the blending of the new and the old is not always a smooth process. Often, it seems that anything that upsets the status quo of one portion of the system, whether or not it is a beneficial change, causes unexpected perturbations throughout the entire system. Sometimes, these unexpected factors can be a surprise bonus in that additional benefits are realized due to unrecognized system interactions. However, in other instances, a modification to the manufacturing system can produce side effects that are unexpected and unwanted.

An example of one unpleasant integration surprise that can happen is the purchase of a system that is incomplete due to a lack of communication of both the buyer's requirements and the capabilities of the vendor. The buyer makes certain assumptions concerning the equipment, based on previous experience, and believes that the package defined by his specification will meet all of his requirements. However, upon installation it is discovered that some item was neglected in the original equipment specification either through a lack of understanding of the standard product or merely due to an oversight. In either case, a problem exists that should have been avoided.

One example of an integration difficulty, that is fairly typical of the type of things that can occur, is demonstrated by a problem that arose during the procurement of an automated FMS for turning operations. Unfortunately, in the preparation of a specification for the automated FMS no criteria were included for a postprocessor to be used in the preparation of part-programs. Since the purchaser neglected to specify the inclusion of this necessary software in the original contract negotiation the vendor did not mention that it was not a part of the system for which a price was quoted. When it was discovered that the vendor did not have the postprocessor and was preparing the programs manually for the system demonstrations, it was too late for the buyer to object. The vendor was meeting the requirements stated in the original specification and the result was an unexpected additional expense for the buyer to acquire the needed capability. Similar events happen in the purchase of software. Often the product literature may cause someone to assume that a feature exists when it really is not available. Another difficulty that can be encountered is promotional material that describes a product feature that is planned for future release but which has not been implemented yet.

One instance of an unexpected benefit that may be received occurs when a system is purchased initially only for one purpose. However, after the equipment has been installed it is determined that there are additional capabilities or applications that were not recognized as being particularly useful during the initial planning phases. An example of this type of circumstance is demonstrated by the introduction of a coordinate measuring machine (CMM) into a production machine shop. In many instances, machine shop personnel are dependent upon manual in-process measurements for determining the machine offsets and for the timely evaluation of the quality of the machining operations. The expected advantage of improved measurement accuracy is readily achieved through the installation and use of the CMM. However, an additional benefit is obtainable through the use of the CMM to provide dimensional feedback information for use with statistical control charts. In a precision manufacturing process, it is not unusual for a manual measurement operation to introduce a relatively large amount of noise into the process data. This excessive variation is due to the manual nature of the operation and precludes the use of this information for obtaining a meaningful statistical analysis. However, the use of the CMM provides dimensional data that is appropriate for use in an accurate assessment of the process quality.

The use of new equipment to discover previously unrecognized characteristics of a manufacturing process is another benefit that can be encountered. The cutting tool vibration-monitoring system, described in an earlier chapter, is another instance in which the integration of a new process monitoring capability provided benefits that exceeded the original expectations. The introduction of this monitoring system into an existing machining operation resulted in the unexpected discovery of a previously unrecognized aspect of the manufacturing operation's characteristics. Historically, the output workpieces from a turning operation had exhibited good dimensional quality. However, there also were occasional periods when the process quality would be degraded, although no one had been able to identify the cause of these intermittent excursions. Excessive tool wear had not been a historical problem. Also, dimensional spot checks of a few readily accessible features on the parts, using the on-machine gaging capability, reinforced this assumption. In addition, a new cutting tool was being used to finish machine each workpiece. Fortunately, the vibration-measurement equipment was available for use as an in-process monitoring system throughout the complete machining cycle. This process monitoring activity identified the difficulty as an intermittent, localized tool-edge-fracture problem. This process anomaly occurred only over a limited portion of the cutting tool and had not been previously suspected because a particular cutting tool was only used to finish machine one part. Hindsight also indicated the need for a more complete series of on-machine measurements as a means of characterizing the quality of the final workpiece.

The previous measurements were not inaccurate. However, they were not extensive enough to be able to provide a complete description of the product quality.

One common question that is encountered in dealing with facility integration is whether to pursue "top down" or "bottom up" implementation of the systems [2]. Sometimes, there is even the perspective that all that is really required is the application of sophisticated computer resources to the planning and scheduling portions of the manufacturing environment. However, while overall planning and design activities may need to proceed from the top down, the physical integration and "closing of control loops" should generally begin at the shop floor and work upward. (It is difficult to justify implementing a DNC system when the basic machines have not even been automated. In addition, it certainly is not practical to attempt to work in either direction without a detailed definition of the necessary interactions and interconnections for the entire system.)

As the control loops associated with the lowest levels are closed the machines become automated and able to function in a semi-independent fashion. This provides results that are readily visible as well as an opportunity to gain acceptance for the working portion of the system from the area personnel. Since there are significant immediate benefits to be gained from early automation of the shop floor equipment it would not be appropriate to perform this portion of the system implementation last. At the same time, there may be multiple uses in a given manufacturing facility for a host computer so that it also can make sense to have this resource available as soon as possible. Therefore, rather than saying that the order of equipment implementation can only occur in a strict fashion, it is more appropriate to consider the entire task and formulate the integration activities in the suitable serial and parallel modes so that the maximum benefit is gained from the effort.

Another facility integration issue that needs to receive a considerable amount of attention is equipment maintenance and repair service. In addition to providing the required support resources for existing equipment, it is also necessary to extend the available maintenance capabilities to provide training and support on new equipment and processes. Another complication is that the tendency is for new systems to be increasingly complex and sometimes more difficult to maintain, although computers have provided the means for sophisticated diagnostic tools to assist the service personnel. If adequate local maintenance facilities are not available then provisions must be made for timely service from external sources whenever it is required. All manufacturing systems eventually need maintenance and sometimes extensive repairs are required. In order to avoid having losses in productivity, system downtime should be minimized whenever possible.

Material flow and part handling are two important facets of facility integration that can be critical bottle necks if they are not handled efficiently. Material that is idle in storage or in a queue is an inefficiently utilized asset. The floor space and other facilities that are associated with this idle material represent additional wasted resources if there is an appropriate alternate use for these areas. Unless adequate facilities exist to get parts to and from a system the usefullness of an advanced manufacturing capability may be wasted. Seemingly minor items such as the number of machines served by an overhead crane can become major stumbling blocks. While one crane per shop may be sufficient for a particular group of production parts, another group that requires significantly less machining time per part may result in wasted machine utilization while the machines wait to be loaded or unloaded. Another similar issue concerns the distribution of cutting tools. Without the required cutting tool, a machining system can be "all fired up with no place to go."

An example of the implementation of an integrated manufacturing philosophy is the continuous flow manufacturing (CFM) concept being utilized by IBM in Rochester, New York [3]. Continuous flow is based on the Japanese kaban approach and involves "building only what you need." This technique keeps work in progress to a minimum because each area supplies only those parts or assemblies that are needed immediately by the next downstream process. Since parts are not produced in large batches the quality of a process becomes very visible. Stockpiling of products is not permitted. Therefore, it is necessary to solve quality problems rather than simply producing a large number of faulty products in order to obtain the necessary number of acceptable workpieces.

IBM began its CFM program with two pilot projects. The initial activity involved a machining line for a hard disk drive mechanism associated with a computer data storage device. A computer simulation predicted that the manufacturing cycle time theoretically could be reduced from 17 days to 15 hours. This seemed to be excessively optimistic but a hand simulation of the operations verified the results of the simulation program. The simulation showed that there was too much inventory which slowed throughput and obscured existing quality problems. When the quality problems were eliminated and the process was tailored to a CFM mode of operation, the cycle time was reduced to 17 hours.

The second pilot project involved a product that depended on a substantial amount of purchased materials. In order to support the CFM system it was necessary to locate outside vendors who would provide frequent, direct delivery of parts and assemblies. IBM worked directly with these suppliers to establish a long-term relationship in which the suppliers also employed CFM techniques within their own facilities.

A final factor that must be considered in the integration of a manufacturing facility is the personnel that will interact with the system. Until factories are run competely without people, there will still be a need for manual involvement with various portions of a facility. Whether the manual intervention is required to operate machines or only limited to maintenance tasks, the integration of these activities into the overall manufacturing system must be considered carefully.

Manual manufacturing systems are notoriously susceptible to operator influence. If a manual system is introduced into an operation without the consideration of the human that must interact with it then problems will arise. While everyone has heard of instances of employees "cutting air instead of making chips," much of the time this problem can be traced to the degree of difficulty encountered in performing the necessary system operations. In general, employees take pride in their skills and the products of their efforts so that it is advantageous to provide them with as much support as practical in performing their tasks. Unfortunately, when problems occur in the manufacturing area, it is often the employees who are cited as the cause. However, the situation may be more involved than it appears at first glance.

Even automatic manufacturing systems usually require manual intervention at some point in the processing operation and these interfaces can become choke points if they are not handled carefully. One good way of avoiding this problem is to have the person who is designing the system spend sufficient time on the shop floor so that the real world problems are understood. Obviously, there must always be some step or factor that paces the manufacturing system, but expensive machinery should not be kept waiting by an inexpensive part of the system. Instead, the manufacturing operation should be paced by the demand for the system output, whenever possible, so that the resources of the facility can be utilized in the most suitable manner.

Summary

Successfully implementing the system integration process for a manufacturing operation involves "bringing it all together" into a viable package that is characterized by smooth interactions with the outside world as well as within its own boundaries. This means that the activities of the various subsystems are blended together to permit problem-free interactions within the system as well as with the outside world. Unfortunately, in some instances project efforts can become too focused on the internal system functions. While the completeness of this area is vital to the success of the eventual system, it is not the only task that needs attention. The result can be that the interfaces that are required to permit efficient operation with the rest of an overall facility are neglected or only considered as an afterthought. In this case, the internal system works well but it is the rest of the manufacturing operation that will be the limiting factor in whether or not the full benefits of a new technology are realized.

A lot of the activities required to achieve the successful integration of a manufacturing system should be guided by the application of common sense in light of the overall goals of the project. However, this does not eliminate the requirement for an intimate understanding of the practical aspects of the various manufacturing operations. Unfortunately, in many projects involving the utilization of new technologies, there are usually some facets that are difficult to predict in advance (except by those having perfect hindsight). Dealing with these "gotchas" requires flexibility and creativity on the part of management as well as the personnel that are responsible for making the system function properly. After all, the new system must be as good or better than the original project description used to justify the expenditure of company resources. Otherwise, it appears that new technologies are being installed just for the sake of having the latest "flashy" system. This position of "technology just for technologies sake" is generally not very easy to defend to someone concerned with operating a profitable enterprise of any type.

An example of the benefits that can be achieved in some instances with relatively little capital investment is typified by a Westinghouse Electric Corporation plant in Ashville, North Carolina. At this facility, a combination of Just-In-Time (JIT) methodology and an objective of utilizing simple solutions were applied to obtain impressive results [4]. This plant achieved cycle time reduction factors ranging from 2 to 4, reduced product warranty costs by 50%, and increased shop productivity by better than 50% while reducing plant inventory by approximately 50%. More that $2 million of capital equipment was eliminated, and 130,000 sq. ft. of manufacturing space was saved. Table 10.1 describes the eight-step process that was employed by Westinghouse personnel to obtain these improvements in their manufacturing processes. While this is a general list of activities that will require different levels of effort at different facilities, there are frequently many things that can be done that require relatively modest resources. One example is the standardization of parts. Originally, Westinghouse was utilizing 265 different crimp terminals throughout its different product lines. After questioning the need for this diversity, it was discovered that 65 types were sufficient. While many of the items on the list are well known as ways of improving a manufacturing operation, it requires the effort to integrate all of these techniques together throughout an operation to obtain the most significant improvements in performance.

Table 10.1 Steps Utilized by Westinghouse Electric Corporation To Streamline the Ashville, N.C. Plant Operations.

1. Create focused mini-factories within the main plant.
2. Streamline order flow.
3. Introduce good manufacturing techniques.
4. Improve material availability.
5. Standardize information, designs, processes, and equipment.
6. Make each employee responsible for quality.
7. Improve customer relationships.
8. Improve communication and training.

The application of sound statistical procedures, especially during the initial justification phases of a project, will provide a firm basis for proceeding with the endeavor. This process information describes the operating characteristics of the existing facility and aids the eventual integration activities. However, an added benefit associated with this task is that gathering and analyzing this data frequently will provide previously unrecognized information. The individuals involved in this activity also gain an extensive insight into the actual condition of existing processes and equipment that would be otherwise difficult to obtain. In addition, the careful use of statistics before, during and after the integration of a manufacturing system provides a firm basis for making accurate comparisons at the completion of the project. Finally, this approach also enhances the performance of the final system because the decisions that are made throughout the project are based on facts rather than on personal opinion. The depth of understanding of the process that is available in the design phases is a critical factor. The completeness and reliability of this information reduces the chances of encountering unpleasant surprises at a later date. It is definitely preferable to have the right design to begin with, rather than having to make extensive system modifications during the implementation process.

References

1. J. F. Manji, Automation at GM assembly plants boosts quality, cuts costs, *Production Engineering, 33*:12, pp. 16–19 (1986).
2. G. J. Blickley, Cell, area, and plant control starts at the bottom, *Control Engineering, 34*:5, pp. 52–54 (1987).
3. Integrated manufacturing: nothing succeeds like successful implementation, *Production Engineering, 34*:5, pp. IM8-IM12 (1987).
4. V. K. Kapoor, Converting to JIT, one plant's experience, *Production Engineering, 34*:2, pp. 42–47 (1987).

CHAPTER 11

Factory of the Future

Introduction

The factory of the future (FOF), paperless factories, robotics, computer control, just-in-time manufacturing, group technology, unattended manufacturing, and so on—all of these topics are being presented at seminars, conferences and other technical presentations as the "wave of the future" which will salvage the manufacturing community from oblivion. Even though there may be a considerable amount of hype in some of these claims, there are facilities at which these technologies are being implemented. At the same time, the other end of the technology/automation spectrum is often a better descriptor of the state of much of the manufacturing community. There are still a lot of manufacturing operations in existence for which NC is just beginning to be benefically applied. Of course, a common road block on the path to achieving CIM in every shop is that the technical and financial resources available to usher in the factory of the future varies widely among different installations. In addition, management is often reluctant to allocate relatively large expenditures for something which varies so drastically from the historical method of operation.

In some instances, it is necessary for the needed changes in the manufacturing processes to occur in an incremental fashion. The advantage of proceeding in this manner is that a system of "learning to walk before attempting to run" can be utilized to demonstrate the validity of the approach. In addition, this method of implementation allows the expenses to be spread out over an extended time period. In other circumstances, the maturity of the technology and the economics of the situation make it much more desirable to accomplish a complete changeover all at once. Irregardless of personal feelings about these changes, the factory of the future will have an impact on everyone, although the results will not be the same for all installations nor should the effects even be similar in many respects. The needs of different manufacturing operations vary greatly and while a full-blown automated FMS may be appropriate in one instance it could be absurd in others. What is identical for all facilities is the need for a common sense approach that is supported with sound statistical methodology.

One characteristic that will be a common denominator among the successful factories of the future, whether they are large or small, simple or complex, is the dependence on repeatability of results. Stability of operations, which means control over the process variability, is required to maintain the quality levels that are needed to establish and retain an effective position in a competitive marketplace. In some instances, this will mean the removal of human operators from certain aspects of the manufacturing process, while in other situations the operations will become completely automated (except for maintenance and other similar activities). For those phases of a process that continue to utilize manual activities, the operations will be designed to permit the work to be accomplished in an efficient and cost effective manner while taking full advantage of the inherent abilities of the workers. In addition, a tight control of the variability of the processes will be achieved while switching from one product to another. Because future lot sizes will be significantly smaller, it will not be acceptable to depend on the conventional technique of spending a certain amount of time "stabilizing a process" at the beginning of a large production run. Process stability will be required during change over operations as well as after the production run is underway. A production run will be considered to be the process of making whatever parts are required over a period of time, not just the time spent making identical products.

In addition to the equipment changes in the manufacturing operations, there also will be alterations in the manner in which facilities are run. One example is the just in time JIT philosophy, mentioned briefly in an earlier chapter, in which the goal is to produce small quantities of a product as they are needed rather than making large weekly or monthly production runs [1]. Just-in-time is a strategy for achieving significant, continuous improvement in the performance of operations by the elimination of wasted time and resources in the complete business activity [2]. The objective is to be continuously involved in activities that prevent waste, improve quality, reduce lead times, and improve people skills and morale. Avoiding waste requires the elimination of problems such as equipment breakdowns, large inventories, products not defined with production requirements in mind, late deliveries, poor quality, poor production planning, and long machine setup times. These problems are solved by attacking them with common sense, not just by spending large sums of money on new equipment.

One easily recognized aspect of maintaining control over the flow of material through the factory is that it is not necessary to maintain large inventories of work in progress or work waiting to be shipped. However, another facet of this manufacturing approach that may not be obvious is the requirement for many setups, equipment changeovers and optimized material handling procedures. For this system to work well, the time required for these activities must be kept to a minimum. The Greeley Division of Hewlett Packard (GLD) is an example of a manufacturing facility in which the implementation of this operating technique was accomplished successfully [3]. Initially, personnel at GLD were pursuing the automation of printed circuit board loading as a means of improving the operation of the facility. However, as the existing process was analyzed it was discovered that only one third of the total labor cost was involved in loading components onto circuit boards. The remainder of the manufacturing costs were due to incoming inspection, repackaging, counting, sorting, testing, and so on. A one cent part could be handled by as many as seven people in the process of being counted three times, placed in four different containers and being involved in four different information generation transactions.

Automation of the printed circuit board loading at GLD could have been accomplished successfully using commercially available systems. However, to choose this direction for the investment of process improvement resources would have dictated ignoring the more fruitful area of activity that was associated with the indirect manufacturing expenses. At this facility, the introduction of the JIT approach to manufacturing was a comprehensive change for the entire organization but the result was well worthwhile. With the old method of operation, the indirect costs associated with the printed circuit board loading activities were twice as high as the direct costs. Since the modification of the mode of operations, the indirect expenses have become one-third as large as the direct costs. While this may seem to be an extreme example, it is actually characteristic of many facilities in which the reasons for doing things become lost in tradition. Too often it seems that common sense is not allowed to prevail in existing operations until it is given a new name and becomes associated with the latest manufacturing "buzz words." The factory of the future is not a magical set of hardware and software that is just around the next corner. Rather, it is a concept or philosophy for doing business in which processes are continually evolving to meet the changing requirements of the marketplace. Usually, it is achieved by making step by step improvements rather than arriving at a temporary utopia in one giant step.

This chapter discusses some of the relatively complex building blocks that are associated with the more complex end of the factory of the future spectrum. The topics covered include manufacturing cells (machining, gaging, assembly, etc.), workpiece and fixture handling systems, fault tolerance approaches, and FMSs. Other elements of the factory of the future such as sensors, statistical process control techniques, computer networks and numerical control systems, and so on may be mentioned briefly. However, detailed information concerning these topics has been presented in earlier chapters.

Manufacturing Cells

In terms of complexity and capability, manufacturing cells may be thought of as existing somewhere between machining centers and FMSs. Machining centers are multifunction, computer controlled machines that are widely used throughout industry. Machining centers often have automatic workpiece gaging and handling features as well as automatic status monitoring and fault reporting capabilities. However, while turning, milling, drilling, and other operations all may be performed on a single machine, it is not considered to be a manufacturing cell because only one machine is involved. (The machining center can be a component in a manufacturing cell or FMS). In contrast, FMS installations represent the current upper end of the manufacturing technology ladder. These systems are described as offering 50% reductions in lead and machining times while requiring only a few machines and operators [4]. While aerospace companies, the defense industry, and many machine tool builders have already installed sophisticated FMSs, other companies are reconsidering their manufacturing requirements in light of the expenses and uncertainties associated with a complete FMS. Vought Aerospace in Dallas, Texas has been operating an FMS since 1984 which is said to be capable of economically producing lot sizes of one unit. However, the not-so-successful FMS purchased by Deere & Company for the Waterloo, Idaho tractor-making factory presents another view of the possible situations that may be encountered.

In general, a manufacturing cell is a scaled down version of a FMS. It may contain machines that are used for metal cutting, gaging, assembly, and so on, as well as material handling equipment and a cell controller. In addition, cells are less complex, less expensive and easier to control than a FMS. Also, manufacturing cells offer some of the same advantages that are associated with a FMS. Attractive features that are available with cells include improved machine utilization, as well as operational flexibility and productivity. These attributes are combined with reduced labor requirements to provide an incremental mechanism for accomplishing the task of complete automation at a later point in time. This can be an attractive method of pursuing the factory of the future, especially for many smaller companies. Due to the advantages associated with these manufacturing cells, it has been estimated that the number of cells installed in the United States will increase from 525 in 1984 to more than 8,000 by 1989 [5].

It is not unusual for manufacturing cells to be available in an almost off-the-shelf fashion from a variety of vendors. A typical configuration that is available would consist of two machining centers and a common pallet system that permits efficient material handling. In addition, the material handling system would support the selection and verification of the appropriate machining part programs and cutting tools based on information that is encoded on the pallet. Several types of tool and pallet identification systems are available [6]. Bar coding is a common technique, which uses a printable, machine-readable language of black bars and white spaces. A scanner or laser reader translates this pattern of bars and spaces into electrical pulses that are decoded to provide a positive identification of each element. Alternatives to the bar code systems include the use of small memory chips that are embedded in the appropriate hardware or radio frequency (RF) transmitters that communicate the identity of the hardware to the cell material management system.

In some cases, parts may leave the initial cell and continue on to another type of cell for further operations. Figure 11.1 shows a diagram of a cell for turning metal billets in preparation for a subsequent series of forming operations. Figure 11.2 shows one of these billets, which has been machined on one end, being loaded onto the machine spindle by a robot which is one element of the billet-cell material handling system. Figure 11.3 shows the other major part of the material handling system which is a conveyor mechanism that is used to supply parts to the machines. Following the forming steps which change the geometric configuration of the workpieces, the parts are transferred to another turning cell in which the rough-formed machining blanks are semifinish machined. Figure 11.4 shows how another robot is utilized in the semifinish machining cell to transfer parts between two machines and the simple conveyor that acts as an input/output manipulator for the workpieces.

Another example of a robotic workcell application is the production of the main engines for the space shuttle [7]. At the Production Enhancement Facility at the Marshall Space Flight Center in Huntsville, Alabama, engineers have developed automated welding processes and tooling for robotic workcells that perform gas tungsten arc welds. The development workcell used to refine the welding process includes a five-axis robot to move the torch along the seam and a two-axis positioner that locates the part in the correct position. The seven axes of motion provide the required freedom of movement to achieve the necessary welding torch/part position at all times. A weld controller maintains the proper voltage, current and gas flow while the weld is tracked using a "through-the-torch" vision system. A CAD/CAM system is employed to develop the motion paths for the robot and part positioning system.

Simplifying material handling requirements also can contribute to the success of a manufacturing cell. An example is the oil pump manufacturing operations conducted by Lufkin Industries, Inc., in Lufkin, Texas [8]. This facility operates in a buyer's market in which it is necessary to respond to a customer's order within 24 hours. Lufkin employs a machining cell with four NC horizontal machining centers to produce the Lufkin pumping unit. The key to Lufkin's success with the new equipment is the reduction in material handling operations that are now necessary. The previous manufacturing method required 15 fixturing setups while the current technique only requires four setups. This simplification of operations combined with the increased accuracies of the new machining centers has allowed product fabrication times to be reduced by a factor of two, while labor costs have been reduced to 25% of the previous value.

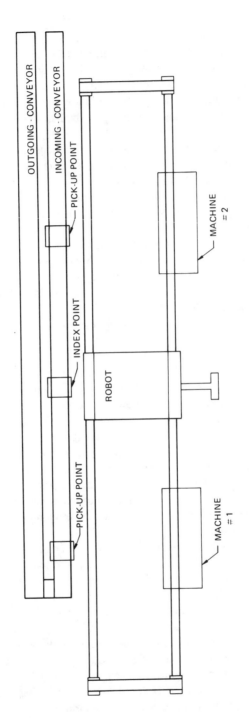

Figure 11.1 Cell for machining metal billets (Courtesy of Martin Marietta Energy Systems, Inc.).

Figure 11.2   Lathe with automatic part handling robot (Courtesy of Martin Marietta Energy Systems, Inc.).

An attractive feature of cells, from a labor expense standpoint, is that one person can usually operate more than one machine. However, this mode of operation means that the individual machines must be capable of performing tasks that were previously accomplished by the person who was running the machine. One example of this type of requirement is a relatively simple housekeeping task. In many cases, it is necessary to have an automatic mechanism to discard the chips that are generated during the machining operations. If this waste material is not controlled it will accumulate and cause serious problems. If these accumulated chips are not kept away from the moving parts of the machine it is likely that some mechanism will become jammed which will bring production on that machine to a rapid halt.

One approach to accomplishing the chip control job is to utilize a set of cutting conditions and or mechanical chip breakers to fracture the chips into small pieces before flushing them away with a stream of coolant. Another technique is to direct a high pressure stream of coolant at the chip as it is being formed so that it is

Figure 11.3 Workpiece conveyor for billet cell (Courtesy of Martin Marietta Energy Systems, Inc.).

broken into fine particles. These broken chips can be so small that they have a sandlike appearance. Then, chip removal from the machine tool can be accomplished by flushing this loose material away with a fluid stream. However, it is necessary to design the machine configuration so that these fines do not become trapped in critical areas such as the slide ways.

Another traditional operator function that must be automated in a manufacturing cell is the detection of circumstances that require an emergency stop condition. Examples of these abnormal conditions include broken cutting tools, machine malfunctions or wrecks and loss of coolant. Since the eyes, ears and intuition of the manufacturing cell operator will not be directed at a particular machine at all times, it is necessary to provide automatic systems that serve the same purpose. (These sensing systems also can be useful for supplementing the abilities of the operator in more conventional manufacturing situations.) The sensors and monitoring systems that can be utilized to accomplish these tasks are available as described in earlier chapters. However, this technology may not be routinely included

Figure 11.4   Cell for semifinish turning hemishell workpieces (Courtesy of Martin Marietta Energy Systems, Inc.).

in a particular manufacturing cell. In this case, it is necessary that the user accept the responsibility of examining the given application to determine what additional sensor systems are required. This activity is the only way of assuring that the necessary equipment and operational features are incorporated into the cell configuration.

In addition to provisions for supplying the cell operator/supervisor with sensor systems for process monitoring, careful consideration must be given to the type of individual that will be depended on to oversee the operation of this manufacturing system. While one of the important objectives of the manufacturing cell may be to reduce the labor requirements associated with a particular operation, it is still probable that a variety of skills will be needed to support this system during daily operations. (In addition, an even wider variety of skills are likely to be required in the event of an equipment malfunction.) This necessity for multiple skills can result in a changing of the traditional job definitions that have tended to be organized around a single craft. Fortunately, in locations such as the General Electric Lynn, Massachusetts plant management and labor have been able to work together creatively to redefine the traditional domains that have been established historically in the manufacturing world. (Unfortunately, in other instances there has been a great deal of resistance to altering the defined responsibilities of different crafts. In these instances, the difficulty is frequently that labor fears a loss of jobs and generally does not trust management.) Usually, the individual that is best suited to perform the job of cell operator/supervisor is a person who has a background in manufacturing. However, it also is very desirable to employ someone who has no fear of dealing with the computer systems (although a healthy degree of distrust can sometimes be justified). In addition, the individual should have an innate curiosity about things as well as a high degree of common sense. In fact, a skilled manufacturing cell operator can listen to the system and detect when there is a problem.

The machine control system requirements associated with a manufacturing cell are only slightly more complex than those needed for an individual machine tool. Since most cells operate on a first in first out (FIFO) basis it is not necessary to employ a control system that can handle any situation at any time. In most cases, once a part enters the cell it always goes through a predictable sequence of operations. This means that the initiation of the next step in the manufacturing sequence is always contingent upon the completion of the previous operation and the task performed by each machine can be determined merely by knowing the identity of the workpiece.

Each of the individual machines in the manufacturing cell has one or more specific part programs that control the machine's response to a given part. These part programs are activated based on the status and location of the workpiece. For example, the robot in Figure 11.4 will load a part from the input conveyor onto the first machine and then wait for a completion signal from the machine control system before executing the actions required to unload the workpiece. In addition, the robot will not attempt to load another part onto the machine even though there may be one waiting at the input station on the conveyor. This occurs because the system recognizes that the machine is busy and unable to accept another part on the spindle.

The control system needed to direct the sequence of activities required by this type of operation is relatively simple. Because the correct action at any time is dictated by the status of the various elements of the system, there is no complicated control algorithm or interpolation-like calculations that must be executed. If an element of the system is busy then the controller just waits until the situation changes. When the status of the cell changes, then the controller executes a decision logic sequence that initiates the performance of the appropriate task. A state table is set up for each workpiece that defines the sequence of processing operations that are required to be completed by the cell. This programming task is relatively uncomplicated and can be implemented easily in an inexpensive PLC system.

Flexible Manufacturing Systems

The definition of a flexible manufacturing system (FMS) is another one of those items that is described differently by many people. In general, it is similar to a manufacturing cell, although that it is significantly more complex. In fact, it may be made up of a number of manufacturing cells that are interconnected by an automated material handling systems. In addition, traditional cost accounting methods are difficult to employ in the justification for the purchase of a FMS. This is because some of the benefits associated with FMSs occur in areas like throughput, quality, responsiveness and so on.

The level of sophistication of the system controller used in a FMS installation is considerably higher than the PLC type of control needed with many cells. For example, while the workpiece routing is fairly constrained in a cell, more processing options exist with a FMS because of the duplication of capabilities. The machining cell shown earlier in Figure 11.4 has two turning machines but they are intended to be utilized to perform two different operations. One station employs a specific type of fixtures for machining the OC of a part while the other station uses different fixtures to machine the IC of the part. While these fixtures can be loaded onto either machine tool, this is a manual operation for this particular system. In a completely automated FMS, the fixtures would be changed through the use of a mechanism that did not require operation intervention. This ability to utilize different machines for the same purpose provides additional routing flexibility that must be controlled through a scheduling system.

In order to manage a manufacturing operation that employs a FMS, a process plan is utilized that defines the steps that are required to fabricate and inspect the workpiece, the order in which these steps must occur, and any alternate sequences of work stations that are acceptable if one or more stations is unavailable at a given time. Currently, the principal approach utilized for FMS dynamic routing is to make the scheduling change decision when a needed part, fixture, or machine becomes available [9]. While the future trend will be to incorporate increasing degrees of "look ahead" capability into the scheduling systems, it is also necessary to consider the economics of developing this capability. The additional gains may not justify an extensive effort at a particular point in the evolution of a given system.

As mentioned above, a significant difference between a manufacturing cell and a FMS is the degree of automation. In a cell system certain operations such as cleaning, lapping and material handling may be performed with various degrees of manual labor. In a completely automated FMS installation, all of these activities are conducted without the assistance of an operator. Of these secondary activities, the material handling task is frequently one of the most difficult to automate. Sometimes, operations which a human executes with relatively little difficulty become quite complex when the person is removed from the process.

Currently, many FMSs utilize automatic guided vehicles (AGVs) or tow-chain carts to transport material between input and output stations in the FMS. The control of these material transporters is frequently handled by a micro or minicomputer. The dispatching of these vehicles must be performed in a logical manner so that neither extensive wait times nor excessive requirements for transporters are encountered. An issue in the planning of the software that controls the dispatching of these transporters is whether or not a cart that is currently in use should be planned for a subsequent task. Allowing all vehicles to be considered for a particular task can result in savings in travel distances and times, while reducing the cart investment and minimizing the transportation delays.

Tool management is another area that becomes increasingly more complex when moving from discrete machines to manufacturing cells to FMS installation. Brother Industries, Limited, Nagoya, Japan, has utilized a FMS to replace a six-year-old DNC system that was used for the production of special heads for industrial sewing machines [10]. This system consists of 13 computer-controlled machining centers that are divided into three lines. The FMS, which is designed to work in production lots of one unit, requires three operators in contrast to the 28 that were needed for the previous DNC line and is heavily dependent on effective tool management. This total management system utilizes a computer-aided inventory approach that tracks warehouse storage and tool movement to the machining centers. An automatic reservation system calculates the proper lead time for the needed tools and automatically delivers them to the FMS after they have been checked for correctness of length and diameter. In use, the tools are monitored for time in use and removed when a job is finished or replaced after a predefined number of operations have been completed.

Because of the complexity and expense of FMS installations, it is desirable to be able to simulate the system during the design phase as a means of evaluating its performance under various sets of operating conditions. Computer simulation is an analytical technique that permits complicated manufacturing facilities to be analyzed without having to construct physical hardware. Most simulation packages offer a visual output that provides an animated description of the performance of the system being studied. The user can readily see problems such as bottlenecks in a process by observing parts piling up at a specific location in an operation. Simulation systems also significantly reduce the risk of instituting production changes in existing systems by offering a preview of what is likely to occur under a particular circumstance.

A side benefit of a simulation system is that it forces the user to carefully analyze the entire manufacturing process in order to achieve an accurate process model. At times, this exercise will provide previously unrecognized insights into a manufacturing operation that can lead to immediate improvements without requiring massive hardware modifications. However, since the data derived from a simulation study is only valid with respect to a specific system model, which is usually only an approximation of the real world, it is necessary to proceed with caution in implementing changes based on simulation results. Whenever possible, it is prudent to perform sufficient hardware testing to verify the results of an analytical study. While it may not be possible to build a prototype of a complete FMS during the early phases of a project, at least the main features of the sytem should be evaluated using actual machinery.

Material Handling

Material handling activities involve the manipulation of a variety of hardware items such as workpieces, tools, fixtures, inspection devices and so on. One common technique for material handling utilizes a transfer line. Although this device is not as flexible in its activities as a fixed robot or AGV, it is successfully employed to transport items throughout a facility. In addition, it can be utilized as an element in the overall automation process. Robots and other programmable material handling systems are capable of more complex motions than a transfer line. However, the best overall system can be one which uses a combination of handling techniques in the most appropriate manner. Fixed robots are suitable for short distances but have a limited reach. Robot vehicles have an extended travel range but they are unavailable once they have accepted a material transfer activity. Transfer lines are usually available and have a long range but they essentially are limited to straight line activities.

Material handling within a cell is typically accomplished through a combination of manipulating devices such as the robots and conveyors shown in Figures 11.2 and 11.4. Robots also may be used to change cutting tools but special purpose mechanisms (often combined with a turret) are frequently employed for this task. Tool changers often consist of a swing arm, double arm, or other automatic changing device that is equipped with the necessary grippers to allow the tool to be transferred between the tool holder and a storage magazine, drum, carousel or rack. In addition, complete tool holders or rotary tools may be changed from a pallet conveyor. Material may be presented to the input of the cell through the use of manual loading techniques or through the use of an automated system such as an AGV.

Material handling systems in a FOF-type facility are significantly different from those that exist in most of today's manufacturing plants. The primary reason for this difference is that the majority of the current manufacturing operations utilize only a small amount (if any) of FOF technology. Existing facilities are largely based on manual material handling methods and depend on human intervention to drive fork lifts, load parts onto machines, acquire tooling from a tool crib, and so on. Factory of the future operations are automated to a significant degree and it is not necessarily appropriate to depend on human actions to perform material handling tasks. The inherent abilities of manufacturing personnel would be utilized in a better fashion than just acting as material transportation systems to move things around.

Another part of the material handling task is the storage and retrieval of the hardware that is needed to make the manufacturing system function without interruption. In the past, materials handling systems were often purchased primarily to improve the material storage activities. However, it is not unusual to find that some material sits in inventory for weeks and months before it is utilized on the shop floor. While there are some critical materials which must be stock piled, in most cases excessive inventory is an unnecessary expense. Modern materials handling systems are more oriented toward the transportation portion of the material handling task (as opposed to the storage problem) so that equipment is moved rapidly throughout the factory.

Automatic storage and retrieval (AS/R) systems are available to organize and control the flow of materials so that they arrive where they are needed in a timely fashion. These systems vary in complexity and capability so that one type system may employ automatic part handling while another utilizes some manual operations. Other differences exist in the types of peripheral equipment that is used. However, the common features that are available include the preparation of kits for assembly, work-in-progress storage, timely delivery of materials and the generation of management/control information.

One example of an automated storage and retrieval facility is a Westinghouse Corporation factory at College Station, Texas [11]. At this location, one manufacturing cell is a standard electronic assembly system and work center that combines material accountability with robotic kitting of components. The automatic storage and retrievel system receives incoming parts, electronically and mechanically inspects and verifies the items, orients the products as required, and inserts them into standard containers. Then, the system automatically preps and kits one carrier for each printing wiring assembly that is to be produced. Each assembly has 500-700 individual components and limited production runs of 10-50 units are frequently encountered.

A second example of a successful AS/R system is at Lockheed, Georgia [12]. In this instance, the AS/R system assists in the automation of the work flow process for parts that will be used in manufacturing aircraft. Jobs consisting of workpieces up to 28 in. long arrive at a cost center where they are logged into the AS/R system and placed into the appropriate bar-coded tote pans. A computer establishes the priority of each job based on the date the part is due for assembly, the number of work centers required and the time needed to complete the work. Then the system retrieves the stored tools so that they can be placed in the tote pan with the part that is to be machined. At this point, the job is released into the workcenter and placed into a queue. As the job moves through the queue to a particular work station, across the machine and on to the next station the system monitors and directs each step along the way. At any point in its route the sequencing can be altered to accommodate a need for repair or rework or to adjust to changes in facility status.

Another aspect of conventional material handling systems is the paper documentation trail that is created as products flow through a manufacturing facility. The proper handling of all this paper can become a monumental task. One example of how an automated material handling system can alleviate this problem is demonstrated by the automated manufacturing factory at the Xerox Fremont Center in Fremont, California. This installation is used to fabricate daisy-wheel printers and electronic typewriters [13]. This facility has been able to strike a balance between the operations required of the mechanical systems and the employees. As incoming parts are received at the warehouse, the supplier's documentation is logged into a computer system where a bar code label is generated that will be used to identify the products as they move through the system. This use of computer control has eliminated 800,000 sheets of paper each year and has contributed to a 66% decrease in production time. The only paper that is generated in the entire manufacturing process is a fail-safe document that describes the parts, quantities, and destinations so that recovery operations are possible in the event of a computer system shutdown. As each assembler uses parts, the delivered goods are released from inventory, tracked to the work station, and assigned to the corresponding finished product without generating any paper other that the fail-safe document. Thousands of such transactions are recorded daily as the integrated computer system tracks the production operations.

Fixturing is another material handling task that plays a critical role in the factory of the future. As products move through a manufacturing operation, it is necessary to secure or clamp different items in certain configurations in order to accomplish the various tasks that are required to convert the raw material into the finished workpiece. In addition, there are usually features or datums on parts that must be properly located before a given manufacturing operation is initiated. One technique for doing this is to secure the product to a pallet which is used to transport the item throughout the factory. In some instances it may not be possible to utilize the same pallet for all of the operations. In this case, a registration device such as is shown in Figure 11.5 and 11.6 may be used to maintain the part alignment across the different setups. This registration mechanism depends on a mechanical hirth coupling which provides automatic alignment between members containing the matching parts. Figure 11.5 shows one half of the coupling attached to a machine spindle, while Figure 11.6 shows a fixture with the mating half of the coupling. When the fixture is placed on the machine spindle so that the two halves of the coupling are in intimate contact, then automatic radial and axial alignment is obtained.

Fault Tolerance

It should be readily apparent that computer systems are the keystone to obtaining the potential benefits associated with factory of the future manufacturing

Figure 11.5   Fixture registration device attached to machine spindle (Courtesy of Martin Marietta Energy Systems, Inc.).

technology. These computers are responsible for real-time allocation of manufacturing resources, as well as other on-line functions such as data archiving, process evaluation, report generation and quality assurance. However, one factor that must be recognized is that all computer systems eventually fail just like other manufacturing equipment and the tolerance of the system to computer faults is a critical design issue [14]. While it can be argued that computer systems fail less often than other manufacturing hardware, ignoring the possibility of computer failure is an unacceptable mode of operation. This is because the consequences of encountering a computer failure can be devastating if no provisions are included in the overall system design to accommodate the situation. Unfortunately, the other extreme approach of providing absolute tolerance of fault conditions is very expensive. What is needed is a middle ground defensive strategy

Figure 11.6   Fixture registration device attached to workpiece fixture (Courtesy of Martin Marietta Energy Systems, Inc.).

that contains the consequences of serious faults without attempting to assure absolutely failure-proof operations. In addition, it is necessary to recognize that while fault-tolerant computers are commercially available, they do not necessarily yield fault-tolerant systems.

Computer faults can occur in the system hardware or the software. In either event, some of the consequences of these faults can be contained by the distribution of functionality throughout a system. However, recovery from software faults can be much more difficult to achieve than recovery from hardware faults. A passive computer failure such as the loss of a power supply or memory unit can cause a complete system shutdown and is easily detected. Therefore, computer stoppage, per se, is not a significant hazard to the system as long as a controlled shut down is achieved and the recovery operations are implemented within the time constants of the critical system functions. Unfortunately, a nonpassive failure may produce corrupted data for some time before the problem is detected and corrected. An example of this type condition is the generation of slightly incorrect commands that do not cause obvious indications of the malfunction and yet significant damage is caused to the process quality.

Reliable software is not always easy to obtain even when software quality assurance techniques are employed. It has been estimated that well-debugged computer software contains approximately 0.5 to 3 bugs per thousand lines of code [15]. These latent bugs are only discovered when a system is put into operation. Obviously, the effects of these errors depend on the application. In addition, in some instances, such as commercial transaction processing systems, it is estimated that 90% of the software bugs are transient. This means that if the software is re-executed then it will work correctly. Some of these bugs are caused by imperfections in hardware or other transient phenomena/user responses that were unanticipated during software development.

A software fault strategy involves a software architecture that detects errors and restores the system to some level of proper operation or else shuts it down in a safe fashion. In addition, this architecture must also meet the initial design specifications assuming fault-free conditions. The application of software fault-tolerance implies the addition to the baseline code of some form of redundancy. However, the largest single source of software errors is due to faults caused by improper or incorrect specifications. Breaking down the programming task at the specifications level and approaching the development task using proven design methodologies also can reduce or eliminate compromising software failures.

References

1. E. J. Hay, Just-In-Time Production—A Winning Combination of Neglected American Ideas, Proceedings of CIMCOM '85, Los Angeles, California (1985).
2. K. R. Plossl, Computer-integrated manufacturing, *Production Engineering,* *34*:6, (1987).
3. G. Winfield, Just-In-Time Manufacturing: A Case Study, Proceedings of CIMCOM '85, Los Angeles, California (1985).
4. D. Palframan, FMS: too much, too soon, *Manufacturing Engineering, 98*:3 (1987).
5. R. P. Bergstrom, FMS: the drive toward cells, *Manufacturing Engineering, 95*:2 (1985).
6. C. Wick, Turning centers update, *Manufacturing Engineering: 99*:1 (1987).
7. D. Stiles, Engine production soars with blend of CAD/CAM & robotics, *Production Engineering, 34*:9 (1987).

8. D. Palframan, Engine production soars with blend of CAD/CAM & robotics, *Production Engineering, 34*:9 (1987).
9. I. S. C. Johnson and K. J. Sharp, Flexible Manufacturing Systems Software Issues and Guidelines, Charles Stark Draper Laboratory, Inc., BDX-613-3578, Cambridge, Mass. (1986).
10. D.J. Gayman and J. Ribeiro, Meeting production needs with tool management. *Manufacturing Engineering, 99*:3 (1987).
11. R. J. Goosman, AS/RS plugs into automated systems, *Production Engineering, 34*:2 (1987).
12. R. L. Martin, AS/RS: from the warehouse to the factory floor, *Manufacturing Engineering, 99*:3 (1987).
13. B. Nagy, Planned automation enhances efficiency, *Manufacturing Engineering, 99*:2 (1987).
14. A. L. Lopkins, Jr., Fault tolerance—how much is enough?, *Manufacturing Engineering, 99*:2 (1987).
15. B. O. Grey, Making SDI software reliable through fault-tolerant techniques, *Defense Electronics, 19*:8 (1987).

Index